T0336785

# DATA MINING IN TIME SERIES DATABASES

**SERIES IN MACHINE PERCEPTION AND ARTIFICIAL INTELLIGENCE***



---

Vol. 43: Agent Engineering
(Eds. *Jiming Liu, Ning Zhong, Yuan Y. Tang and Patrick S. P. Wang*)

Vol. 44: Multispectral Image Processing and Pattern Recognition
(Eds. *J. Shen, P. S. P. Wang and T. Zhang*)

Vol. 45: Hidden Markov Models: Applications in Computer Vision
(Eds. *H. Bunke and T. Caelli*)

Vol. 46: Syntactic Pattern Recognition for Seismic Oil Exploration
(*K. Y. Huang*)

Vol. 47: Hybrid Methods in Pattern Recognition
(Eds. *H. Bunke and A. Kandel*)

Vol. 48: Multimodal Interface for Human-Machine Communications
(Eds. *P. C. Yuen, Y. Y. Tang and P. S. P. Wang*)

Vol. 49: Neural Networks and Systolic Array Design
(Eds. *D. Zhang and S. K. Pal*)

Vol. 50: Empirical Evaluation Methods in Computer Vision
(Eds. *H. I. Christensen and P. J. Phillips*)

Vol. 51: Automatic Diatom Identification
(Eds. *H. du Buf and M. M. Bayer*)

Vol. 52: Advances in Image Processing and Understanding
A Festschrift for Thomas S. Huwang
(Eds. *A. C. Bovik, C. W. Chen and D. Goldgof*)

Vol. 53: Soft Computing Approach to Pattern Recognition and Image Processing
(Eds. *A. Ghosh and S. K. Pal*)

Vol. 54: Fundamentals of Robotics — Linking Perception to Action
(*M. Xie*)

Vol. 55: Web Document Analysis: Challenges and Opportunities
(Eds. *A. Antonacopoulos and J. Hu*)

Vol. 56: Artificial Intelligence Methods in Software Testing
(Eds. *M. Last, A. Kandel and H. Bunke*)

Vol. 57: Data Mining in Time Series Databases
(Eds. *M. Last, A. Kandel and H. Bunke*)

Vol. 58: Computational Web Intelligence: Intelligent Technology for
Web Applications
(Eds. *Y. Zhang, A. Kandel, T. Y. Lin and Y. Yao*)

Vol. 59: Fuzzy Neural Network Theory and Application
(*P. Liu and H. Li*)

*For the complete list of titles in this series, please write to the Publisher.

Series in Machine Perception and Artificial Intelligence – Vol. 57

# DATA MINING IN TIME SERIES DATABASES

Mark Last
Ben-Gurion University of the Negev, Israel

Abraham Kandel
Tel-Aviv University Israel & University of South Florida, USA

Abraham Kandel
University of Bern, Switzerland

World Scientific

NEW JERSEY • LONDON • SINGAPORE • BEIJING • SHANGHAI • HONG KONG • TAIPEI • CHENNAI

*Published by*

World Scientific Publishing Co. Pte. Ltd.
5 Toh Tuck Link, Singapore 596224
*USA office:* 27 Warren Street, Suite 401-402, Hackensack, NJ 07601
*UK office:* 57 Shelton Street, Covent Garden, London WC2H 9HE

**British Library Cataloguing-in-Publication Data**
A catalogue record for this book is available from the British Library.

First published 2004
Reprinted 2005

**DATA MINING IN TIME SERIES DATABASES**
**Series in Machine Perception and Artificial Intelligence (Vol. 57)**



ISBN-13 978-981-238-290-0
ISBN-10 981-238-290-9

Typeset by Stallion Press
Email: enquiries@stallionpress.com

Printed in Singapore

**Dedicated to**

*The Honorable Congressman C. W. Bill Young*
*House of Representatives*

For his vision and continuous support in creating the National Institute for Systems Test and Productivity at the Computer Science and Engineering Department, University of South Florida

# Preface

Traditional data mining methods are designed to deal with "static" databases, i.e. databases where the *ordering* of records (or other database objects) has nothing to do with the patterns of interest. Though the assumption of order irrelevance may be sufficiently accurate in some applications, there are certainly many other cases, where sequential information, such as a time-stamp associated with every record, can significantly enhance our knowledge about the mined data. One example is a series of stock values: a specific closing price recorded yesterday has a completely different meaning than the same value a year ago. Since most today's databases already include temporal data in the form of "date created", "date modified", and other time-related fields, the only problem is how to exploit this valuable information to our benefit. In other words, the question we are currently facing is: *How to mine time series data?*

The purpose of this volume is to present some recent advances in preprocessing, mining, and interpretation of temporal data that is stored by modern information systems. Adding the time dimension to a database produces a Time Series Database (TSDB) and introduces new aspects and challenges to the tasks of data mining and knowledge discovery. These new challenges include: finding the most efficient representation of time series data, measuring similarity of time series, detecting change points in time series, and time series classification and clustering. Some of these problems have been treated in the past by experts in *time series analysis*. However, statistical methods of time series analysis are focused on sequences of values representing a single numeric variable (e.g., price of a specific stock). In a real-world database, a time-stamped record may include several numerical and nominal attributes, which may depend not only on the time dimension but also on each other. To make the data mining task even more complicated, the objects in a time series may represent some complex graph structures rather than vectors of feature-values.

Our book covers the state-of-the-art research in several areas of time series data mining. Specific problems challenged by the authors of this volume are as follows.

*Representation of Time Series.* Efficient and effective representation of time series is a key to successful discovery of time-related patterns. The most frequently used representation of single-variable time series is piecewise linear approximation, where the original points are reduced to a set of straight lines ("segments"). Chapter 1 by Eamonn Keogh, Selina Chu, David Hart, and Michael Pazzani provides an extensive and comparative overview of existing techniques for time series segmentation. In the view of shortcomings of existing approaches, the same chapter introduces an improved segmentation algorithm called SWAB (Sliding Window and Bottom-up).

*Indexing and Retrieval of Time Series.* Since each time series is characterized by a large, potentially unlimited number of points, finding two *identical* time series for any phenomenon is hopeless. Thus, researchers have been looking for sets of *similar* data sequences that differ only slightly from each other. The problem of retrieving similar series arises in many areas such as marketing and stock data analysis, meteorological studies, and medical diagnosis. An overview of current methods for efficient retrieval of time series is presented in Chapter 2 by Magnus Lie Hetland. Chapter 3 (by Eugene Fink and Kevin B. Pratt) presents a new method for fast compression and indexing of time series. A robust similarity measure for retrieval of noisy time series is described and evaluated by Michail Vlachos, Dimitrios Gunopulos, and Gautam Das in Chapter 4.

*Change Detection in Time Series.* The problem of change point detection in a sequence of values has been studied in the past, especially in the context of time series segmentation (see above). However, the nature of real-world time series may be much more complex, involving multivariate and even graph data. Chapter 5 (by Gil Zeira, Oded Maimon, Mark Last, and Lior Rokach) covers the problem of change detection in a classification model induced by a data mining algorithm from time series data. A change detection procedure for detecting abnormal events in time series of graphs is presented by Horst Bunke and Miro Kraetzl in Chapter 6. The procedure is applied to abnormal event detection in a computer network.

*Classification of Time Series.* Rather than partitioning a time series into segments, one can see each time series, or any other sequence of data points, as a single object. Classification and clustering of such complex "objects" may be particularly beneficial for the areas of process control, intrusion detection, and character recognition. In Chapter 7, Carlos J. Alonso González and Juan J. Rodríguez Diez present a new method for early classification of multivariate time series. Their method is capable of learning from series of variable length and able of providing a classification when only part of the series is presented to the classifier. A novel concept of representing time series by median strings (see Chapter 8, by Xiaoyi Jiang, Horst Bunke, and Janos Csirik) opens new opportunities for applying classification and clustering methods of data mining to sequential data.

As indicated above, the area of mining time series databases still includes many unexplored and insufficiently explored issues. Specific suggestions for future research can be found in individual chapters. In general, we believe that interesting and useful results can be obtained by applying the methods described in this book to real-world sets of sequential data.

**Acknowledgments**

The preparation of this volume was partially supported by the National Institute for Systems Test and Productivity at the University of South Florida under U.S. Space and Naval Warfare Systems Command grant number N00039-01-1-2248.

We also would like to acknowledge the generous support and cooperation of: Ben-Gurion University of the Negev, Department of Information Systems Engineering, University of South Florida, Department of Computer Science and Engineering, Tel-Aviv University, College of Engineering, The Fulbright Foundation, The US-Israel Educational Foundation.

*January 2004*

*Mark Last*
*Abraham Kandel*
*Horst Bunke*

## Contents

# CHAPTER 1

# SEGMENTING TIME SERIES: A SURVEY AND NOVEL APPROACH

Eamonn Keogh

*Computer Science & Engineering Department, University of California — Riverside, Riverside, California 92521, USA*

E-mail: eamonn@cs.ucr.edu

Selina Chu, David Hart, and Michael Pazzani

*Department of Information and Computer Science, University of California, Irvine, California 92697, USA*

E-mail: {selina, dhart, pazzani}@ics.uci.edu

In recent years, there has been an explosion of interest in mining time series databases. As with most computer science problems, representation of the data is the key to efficient and effective solutions. One of the most commonly used representations is piecewise linear approximation. This representation has been used by various researchers to support clustering, classification, indexing and association rule mining of time series data. A variety of algorithms have been proposed to obtain this representation, with several algorithms having been independently rediscovered several times. In this chapter, we undertake the first extensive review and empirical comparison of all proposed techniques. We show that all these algorithms have fatal flaws from a data mining perspective. We introduce a novel algorithm that we empirically show to be superior to all others in the literature.

*Keywords*: Time series; data mining; piecewise linear approximation; segmentation; regression.

## 1. Introduction

In recent years, there has been an explosion of interest in mining time series databases. As with most computer science problems, representation of the data is the key to efficient and effective solutions. Several high level

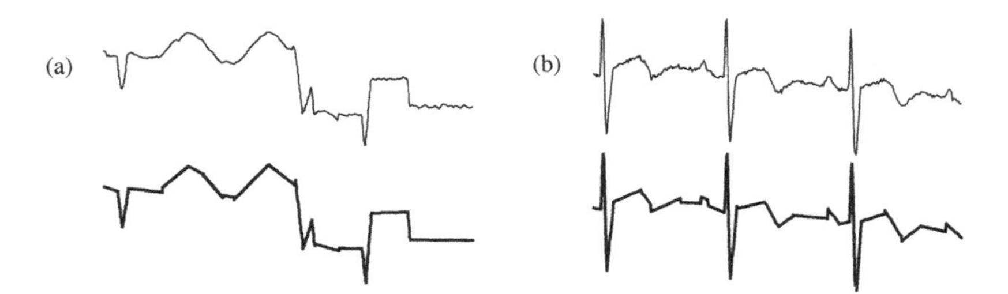


Fig. 1. Two time series and their piecewise linear representation. (a) Space Shuttle Telemetry. (b) Electrocardiogram (ECG).

representations of time series have been proposed, including Fourier Transforms [Agrawal *et al.* (1993), Keogh *et al.* (2000)], Wavelets [Chan and Fu (1999)], Symbolic Mappings [Agrawal *et al.* (1995), Das *et al.* (1998), Perng *et al.* (2000)] and Piecewise Linear Representation (PLR). In this work, we confine our attention to PLR, perhaps the most frequently used representation [Ge and Smyth (2001), Last *et al.* (2001), Hunter and McIntosh (1999), Koski *et al.* (1995), Keogh and Pazzani (1998), Keogh and Pazzani (1999), Keogh and Smyth (1997), Lavrenko *et al.* (2000), Li *et al.* (1998), Osaki *et al.* (1999), Park *et al.* (2001), Park *et al.* (1999), Qu *et al.* (1998), Shatkay (1995), Shatkay and Zdonik (1996), Vullings *et al.* (1997), Wang and Wang (2000)].

Intuitively, Piecewise Linear Representation refers to the approximation of a time series $T$, of length $n$, with K straight lines (hereafter known as segments). Figure 1 contains two examples. Because K is typically much smaller that $n$, this representation makes the storage, transmission and computation of the data more efficient. Specifically, in the context of data mining, the piecewise linear representation has been used to:

- Support fast exact similarly search [Keogh *et al.* (2000)].
- Support novel distance measures for time series, including "fuzzy queries" [Shatkay (1995), Shatkay and Zdonik (1996)], weighted queries [Keogh and Pazzani (1998)], multiresolution queries [Wang and Wang (2000), Li *et al.* (1998)], dynamic time warping [Park *et al.* (1999)] and relevance feedback [Keogh and Pazzani (1999)].
- Support concurrent mining of text and time series [Lavrenko *et al.* (2000)].
- Support novel clustering and classification algorithms [Keogh and Pazzani (1998)].
- Support change point detection [Sugiura and Ogden (1994), Ge and Smyth (2001)].

Surprisingly, in spite of the ubiquity of this representation, with the exception of [Shatkay (1995)], there has been little attempt to understand and compare the algorithms that produce it. Indeed, there does not even appear to be a consensus on what to call such an algorithm. For clarity, we will refer to these types of algorithm, which input a time series and return a piecewise linear representation, as *segmentation* algorithms.

The segmentation problem can be framed in several ways.

- Given a time series $T$, produce the best representation using only K segments.
- Given a time series $T$, produce the best representation such that the maximum error for any segment does not exceed some user-specified threshold, max_error.
- Given a time series $T$, produce the best representation such that the combined error of all segments is less than some user-specified threshold, total_max_error.

As we shall see in later sections, not all algorithms can support all these specifications.

Segmentation algorithms can also be classified as batch or online. This is an important distinction because many data mining problems are inherently dynamic [Vullings *et al.* (1997), Koski *et al.* (1995)].

Data mining researchers, who needed to produce a piecewise linear approximation, have typically either independently rediscovered an algorithm or used an approach suggested in related literature. For example, from the fields of cartography or computer graphics [Douglas and Peucker (1973), Heckbert and Garland (1997), Ramer (1972)].

In this chapter, we review the three major segmentation approaches in the literature and provide an extensive empirical evaluation on a very heterogeneous collection of datasets from finance, medicine, manufacturing and science. The major result of these experiments is that only online algorithm in the literature produces very poor approximations of the data, and that the only algorithm that consistently produces high quality results and scales linearly in the size of the data is a batch algorithm. These results motivated us to introduce a new online algorithm that scales linearly in the size of the data set, is online, and produces high quality approximations.

The rest of the chapter is organized as follows. In Section 2, we provide an extensive review of the algorithms in the literature. We explain the basic approaches, and the various modifications and extensions by data miners. In Section 3, we provide a detailed empirical comparison of all the algorithms.

We will show that the most popular algorithms used by data miners can in fact produce very poor approximations of the data. The results will be used to motivate the need for a new algorithm that we will introduce and validate in Section 4. Section 5 offers conclusions and directions for future work.

## 2. Background and Related Work

In this section, we describe the three major approaches to time series segmentation in detail. Almost all the algorithms have 2 and 3 dimensional analogues, which ironically seem to be better understood. A discussion of the higher dimensional cases is beyond the scope of this chapter. We refer the interested reader to [Heckbert and Garland (1997)], which contains an excellent survey.

Although appearing under different names and with slightly different implementation details, most time series segmentation algorithms can be grouped into one of the following three categories:

- *Sliding Windows:* A segment is grown until it exceeds some error bound. The process repeats with the next data point not included in the newly approximated segment.
- *Top-Down:* The time series is recursively partitioned until some stopping criteria is met.
- *Bottom-Up:* Starting from the finest possible approximation, segments are merged until some stopping criteria is met.

Table 1 contains the notation used in this chapter.

Table 1. Notation.

| | |
|---|---|
| $T$ | A time series in the form $t_1, t_2, \ldots, t_n$ |
| $T[a:b]$ | The subsection of $T$ from $a$ to $b$, $t_a, t_{a+1}, \ldots, t_b$ |
| *Seg_TS* | A piecewise linear approximation of a time series of length $n$ with $K$ segments. Individual segments can be addressed with $Seg\_TS(i)$. |
| *create_segment*($T$) | A function that takes in a time series and returns a linear segment approximation of it. |
| *calculate_error*($T$) | A function that takes in a time series and returns the approximation error of the linear segment approximation of it. |

Given that we are going to approximate a time series with straight lines, there are at least two ways we can find the approximating line.

- *Linear Interpolation:* Here the approximating line for the subsequence $T[a : b]$ is simply the line connecting $t_a$ and $t_b$. This can be obtained in constant time.
- *Linear Regression:* Here the approximating line for the subsequence $T[a : b]$ is taken to be the best fitting line in the least squares sense [Shatkay (1995)]. This can be obtained in time linear in the length of segment.

The two techniques are illustrated in Figure 2. Linear interpolation tends to closely align the endpoint of consecutive segments, giving the piecewise approximation a "smooth" look. In contrast, piecewise linear regression can produce a very disjointed look on some datasets. The aesthetic superiority of linear interpolation, together with its low computational complexity has made it the technique of choice in computer graphic applications [Heckbert and Garland (1997)]. However, the quality of the approximating line, in terms of Euclidean distance, is generally inferior to the regression approach.

In this chapter, we deliberately keep our descriptions of algorithms at a high level, so that either technique can be imagined as the approximation technique. In particular, the pseudocode function `create_segment(T)` can be imagined as using interpolation, regression or any other technique.

All segmentation algorithms also need some method to evaluate the quality of fit for a potential segment. A measure commonly used in conjunction with linear regression is the sum of squares, or the residual error. This is calculated by taking all the vertical differences between the best-fit line and the actual data points, squaring them and then summing them together. Another commonly used measure of goodness of fit is the distance between the best fit line and the data point furthest away in the vertical direction

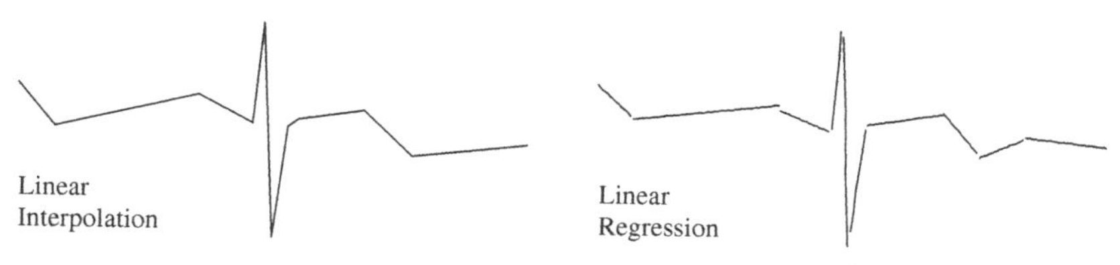


Fig. 2. Two 10-segment approximations of electrocardiogram data. The approximation created using linear interpolation has a smooth aesthetically appealing appearance because all the endpoints of the segments are aligned. Linear regression, in contrast, produces a slightly disjointed appearance but a tighter approximation in terms of residual error.

(i.e. the L$\infty$ norm between the line and the data). As before, we have kept our descriptions of the algorithms general enough to encompass any error measure. In particular, the pseudocode function `calculate_error(T)` can be imagined as using any sum of squares, furthest point, or any other measure.

## 2.1. *The Sliding Window Algorithm*

The Sliding Window algorithm works by anchoring the left point of a potential segment at the first data point of a time series, then attempting to approximate the data to the right with increasing longer segments. At some point $i$, the error for the potential segment is greater than the user-specified threshold, so the subsequence from the anchor to $i - 1$ is transformed into a segment. The anchor is moved to location $i$, and the process repeats until the entire time series has been transformed into a piecewise linear approximation. The pseudocode for the algorithm is shown in Table 2.

The Sliding Window algorithm is attractive because of its great simplicity, intuitiveness and particularly the fact that it is an online algorithm. Several variations and optimizations of the basic algorithm have been proposed. Koski *et al.* noted that on ECG data it is possible to speed up the algorithm by incrementing the variable $i$ by "leaps of length $k$" instead of 1. For $k = 15$ (at 400 Hz), the algorithm is 15 times faster with little effect on the output accuracy [Koski *et al.* (1995)].

Depending on the error measure used, there may be other optimizations possible. Vullings *et al.* noted that since the residual error is monotonically non-decreasing with the addition of more data points, one does not have to test every value of $i$ from 2 to the final chosen value [Vullings *et al.* (1997)]. They suggest initially setting $i$ to $s$, where $s$ is the mean length of the previous segments. If the guess was pessimistic (the measured error

Table 2. The generic Sliding Window algorithm.

```
Algorithm Seg_TS = Sliding_Window(T, max_error)
anchor = 1;
while not finished segmenting time series
i = 2;
  while  calculate_error(T[anchor: anchor + i ]) < max_error
    i = i + 1;
  end;
  Seg_TS = concat(Seg_TS, create_segment(T[anchor: anchor
      + (i - 1)]);anchor = anchor + i;
end;
```

is still less than `max_error`) then the algorithm continues to increment $i$ as in the classic algorithm. Otherwise they begin to decrement $i$ until the measured error is less than `max_error`. This optimization can greatly speed up the algorithm if the mean length of segments is large in relation to the standard deviation of their length. The monotonically non-decreasing property of residual error also allows binary search for the length of the segment. Surprisingly, no one we are aware of has suggested this.

The Sliding Window algorithm can give pathologically poor results under some circumstances, particularly if the time series in question contains abrupt level changes. Most researchers have not reported this [Qu *et al.* (1998), Wang and Wang (2000)], perhaps because they tested the algorithm on stock market data, and its relative performance is best on noisy data. Shatkay (1995), in contrast, does notice the problem and gives elegant examples and explanations [Shatkay (1995)]. They consider three variants of the basic algorithm, each designed to be robust to a certain case, but they underline the difficulty of producing a single variant of the algorithm that is robust to arbitrary data sources.

Park *et al.* (2001) suggested modifying the algorithm to create "*monotonically changing*" segments [Park *et al.* (2001)]. That is, all segments consist of data points of the form of $t_1 \leq t_2 \leq \cdots \leq t_n$ or $t_1 \geq t_2 \geq \cdots \geq t_n$. This modification worked well on the smooth synthetic dataset it was demonstrated on. But on real world datasets with any amount of noise, the approximation is greatly overfragmented.

Variations on the Sliding Window algorithm are particularly popular with the medical community (where it is known as FAN or SAPA), since patient monitoring is inherently an online task [Ishijima *et al.* (1983), Koski *et al.* (1995), McKee *et al.* (1994), Vullings *et al.* (1997)].

## 2.2. *The Top-Down Algorithm*

The Top-Down algorithm works by considering every possible partitioning of the times series and splitting it at the best location. Both subsections are then tested to see if their approximation error is below some user-specified threshold. If not, the algorithm recursively continues to split the subsequences until all the segments have approximation errors below the threshold. The pseudocode for the algorithm is shown in Table 3.

Variations on the Top-Down algorithm (including the 2-dimensional case) were independently introduced in several fields in the early 1970's. In cartography, it is known as the Douglas-Peucker algorithm [Douglas and

Table 3. The generic Top-Down algorithm.

```
Algorithm Seg_TS = Top_Down(T, max_error)
best_so_far = inf;
for i = 2 to length(T) - 2    // Find the best splitting point.
  improvement_in_approximation = improvement_splitting_here(T, i);
  if improvement_in_approximation < best_so_far
     breakpoint = i;
     best_so_far = improvement_in_approximation;
  end;
end;
  // Recursively split the left segment if necessary.
if calculate_error(T[1:breakpoint]) > max_error
  Seg_TS = Top_Down(T[1:breakpoint]);
end;
  // Recursively split the right segment if necessary.
if calculate_error(T[breakpoint + 1:length(T)]) > max_error
  Seg_TS = Top_Down(T[breakpoint + 1:length(T)]);
end;
```

Peucker (1973)]; in image processing, it is known as Ramer's algorithm [Ramer (1972)]. Most researchers in the machine learning/data mining community are introduced to the algorithm in the classic textbook by Duda and Harts, which calls it "*Iterative End-Points Fits*" [Duda and Hart (1973)].

In the data mining community, the algorithm has been used by [Li *et al.* (1998)] to support a framework for mining sequence databases at multiple abstraction levels. Shatkay and Zdonik use it (after considering alternatives such as Sliding Windows) to support approximate queries in time series databases [Shatkay and Zdonik (1996)].

Park *et al.* introduced a modification where they first perform a scan over the entire dataset marking every peak and valley [Park *et al.* (1999)]. These extreme points used to create an initial segmentation, and the Top-Down algorithm is applied to each of the segments (in case the error on an individual segment was still too high). They then use the segmentation to support a special case of dynamic time warping. This modification worked well on the smooth synthetic dataset it was demonstrated on. But on real world data sets with any amount of noise, the approximation is greatly overfragmented.

Lavrenko *et al.* uses the Top-Down algorithm to support the concurrent mining of text and time series [Lavrenko *et al.* (2000)]. They attempt to discover the influence of news stories on financial markets. Their algorithm contains some interesting modifications including a novel stopping criteria based on the t-test.

Finally Smyth and Ge use the algorithm to produce a representation that can support a Hidden Markov Model approach to both change point detection and pattern matching [Ge and Smyth (2001)].

## 2.3. *The Bottom-Up Algorithm*

The Bottom-Up algorithm is the natural complement to the Top-Down algorithm. The algorithm begins by creating the finest possible approximation of the time series, so that $n/2$ segments are used to approximate the $n$-length time series. Next, the cost of merging each pair of adjacent segments is calculated, and the algorithm begins to iteratively merge the lowest cost pair until a stopping criteria is met. When the pair of adjacent segments $i$ and $i + 1$ are merged, the algorithm needs to perform some bookkeeping. First, the cost of merging the new segment with its right neighbor must be calculated. In addition, the cost of merging the $i - 1$ segments with its new larger neighbor must be recalculated. The pseudocode for the algorithm is shown in Table 4.

Two and three-dimensional analogues of this algorithm are common in the field of computer graphics where they are called decimation methods [Heckbert and Garland (1997)]. In data mining, the algorithm has been used extensively by two of the current authors to support a variety of time series data mining tasks [Keogh and Pazzani (1999), Keogh and Pazzani (1998), Keogh and Smyth (1997)]. In medicine, the algorithm was used by Hunter and McIntosh to provide the high level representation for their medical pattern matching system [Hunter and McIntosh (1999)].

Table 4. The generic Bottom-Up algorithm.

```
Algorithm Seg_TS = Bottom_Up(T, max_error)
for i = 1 : 2 : length(T)        // Create initial fine approximation.
  Seg_TS = concat(Seg_TS, create_segment(T[i: i + 1 ]));
end;
for i = 1 : length(Seg_TS) - 1                      // Find merging costs.
  merge_cost(i) = calculate_error([merge(Seg_TS(i), Seg_TS(i + 1))]);
end;
while min(merge_cost) < max_error                    // While not finished.
  p = min(merge_cost);      // Find ''cheapest'' pair to merge.
  Seg_TS(p) = merge(Seg_TS(p), Seg_TS(p + 1));              // Merge them.
  delete(Seg_TS(p + 1));                               // Update records.
  merge_cost(p) = calculate_error(merge(Seg_TS(p), Seg_TS(p + 1)));
  merge_cost(p - 1) = calculate_error(merge(Seg_TS(p - 1), Seg_TS(p)));
end;
```

### 2.4. *Feature Comparison of the Major Algorithms*

We have deliberately deferred the discussion of the running times of the algorithms until now, when the reader's intuition for the various approaches are more developed. The running time for each approach is data dependent. For that reason, we discuss both a worst-case time that gives an upper bound and a best-case time that gives a lower bound for each approach.

We use the standard notation of $\Omega(f(n))$ for a lower bound, $O(f(n))$ for an upper bound, and $\theta(f(n))$ for a function that is both a lower and upper bound.

**Definitions and Assumptions.** The number of data points is $n$, the number of segments we plan to create is $K$, and thus the average segment length is $L = n/K$. The actual length of segments created by an algorithm varies and we will refer to the lengths as $L_i$.

All algorithms, except top-down, perform considerably worse if we allow any of the $L_I$ to become very large (say $n/4$), so we assume that the algorithms limit the maximum length $L$ to some multiple of the average length. It is trivial to code the algorithms to enforce this, so the time analysis that follows is exact when the algorithm includes this limit. Empirical results show, however, that the segments generated (with no limit on length) are tightly clustered around the average length, so this limit has little effect in practice.

We assume that for each set $S$ of points, we compute a best segment and compute the error in $\theta(n)$ time. This reflects the way these algorithms are coded in practice, which is to use a packaged algorithm or function to do linear regression. We note, however, that we believe one can produce asymptotically faster algorithms if one custom codes linear regression (or other best fit algorithms) to reuse computed values so that the computation is done in less than $O(n)$ time in subsequent steps. We leave that as a topic for future work. In what follows, all computations of best segment and error are assumed to be $\theta(n)$.

**Top-Down.** The best time for Top-Down occurs if each split occurs at the midpoint of the data. The first iteration computes, for each split point $i$, the best line for points $[1,\ i]$ and for points $[i+1,\ n]$. This takes $\theta(n)$ for each split point, or $\theta(n^2)$ total for all split points. The next iteration finds split points for $[1,\ n/2]$ and for $[n/2+1,\ n]$. This gives a recurrence $T(n) = 2\mathrm{T}(n/2) + \theta(n^2)$ where we have $T(2) = c$, and this solves to $T(n) = \Omega(n^2)$. This is a lower bound because we assumed the data has the best possible split points.

The worst time occurs if the computed split point is always at one side (leaving just 2 points on one side), rather than the middle. The recurrence is $T(n) = T(n-2) + \theta(n^2)$ We must stop after $K$ iterations, giving a time of $O(n^2K)$.

**Sliding Windows.** For this algorithm, we compute best segments for larger and larger windows, going from 2 up to at most $cL$ (by the assumption we discussed above). The maximum time to compute a single segment is $\sum_{i=2}^{cL} \theta(i) = \theta(L^2)$. The number of segments can be as few as $n/cL = K/c$ or as many as $K$. The time is thus $\theta(L^2K)$ or $\theta(Ln)$. This is both a best case and worst case bound.

**Bottom-Up.** The first iteration computes the segment through each pair of points and the costs of merging adjacent segments. This is easily seen to take $O(n)$ time. In the following iterations, we look up the minimum error pair $i$ and $i+1$ to merge; merge the pair into a new segment $S_{new}$; delete from a heap (keeping track of costs is best done with a heap) the costs of merging segments $i-1$ and $i$ and merging segments $i+1$ and $i+2$; compute the costs of merging $S_{new}$ with $S_{i-1}$ and with $S_{i-2}$; and insert these costs into our heap of costs. The time to look up the best cost is $\theta(1)$ and the time to add and delete costs from the heap is $O(\log n)$. (The time to construct the heap is $O(n)$.)

In the best case, the merged segments always have about equal length, and the final segments have length $L$. The time to merge a set of length 2 segments, which will end up being one length $L$ segment, into half as many segments is $\theta(L)$ (for the time to compute the best segment for every pair of merged segments), not counting heap operations. Each iteration takes the same time repeating $\theta(\log L)$ times gives a segment of size $L$.

The number of times we produce length $L$ segments is $K$, so the total time is $\Omega(K\, L \log L) = \Omega(n \log n/K)$. The heap operations may take as much as $O(n \log n)$. For a lower bound we have proven just $\Omega(n \log n/K)$.

In the worst case, the merges always involve a short and long segment, and the final segments are mostly of length $cL$. The time to compute the cost of merging a length 2 segment with a length $i$ segment is $\theta(i)$, and the time to reach a length $cL$ segment is $\sum_{i=2}^{cL} \theta(i) = \theta(L^2)$. There are at most $n/cL$ such segments to compute, so the time is $n/cL \times \theta(L^2) = O(Ln)$. (Time for heap operations is inconsequential.) This complexity study is summarized in Table 5.

In addition to the time complexity there are other features a practitioner might consider when choosing an algorithm. First there is the question of

Table 5. A feature summary for the 3 major algorithms.

| Algorithm | User can specify[1] | Online | Complexity |
|---|---|---|---|
| Top-Down | $E$, $ME$, $K$ | No | $O(n^2K)$ |
| Bottom-Up | $E$, $ME$, $K$ | No | $O(Ln)$ |
| Sliding Window | $E$ | Yes | $O(Ln)$ |

[1]KEY: $E \rightarrow$ Maximum error for a given segment, $ME \rightarrow$ Maximum error for a given segment for entire time series, $K \rightarrow$ Number of segments.

whether the algorithm is online or batch. Secondly, there is the question of how the user can specify the quality of desired approximation. With trivial modifications the Bottom-Up algorithm allows the user to specify the desired value of $K$, the maximum error per segment, or total error of the approximation. A (non-recursive) implementation of Top-Down can also be made to support all three options. However Sliding Window only allows the maximum error per segment to be specified.

## 3. Empirical Comparison of the Major Segmentation Algorithms

In this section, we will provide an extensive empirical comparison of the three major algorithms. It is possible to create artificial datasets that allow one of the algorithms to achieve zero error (by any measure), but forces the other two approaches to produce arbitrarily poor approximations. In contrast, testing on purely random data forces the all algorithms to produce essentially the same results. To overcome the potential for biased results, we tested the algorithms on a very diverse collection of datasets. These datasets where chosen to represent the extremes along the following dimensions, stationary/non-stationary, noisy/smooth, cyclical/non-cyclical, symmetric/asymmetric, etc. In addition, the data sets represent the diverse areas in which data miners apply their algorithms, including finance, medicine, manufacturing and science. Figure 3 illustrates the 10 datasets used in the experiments.

### 3.1. *Experimental Methodology*

For simplicity and brevity, we only include the linear regression versions of the algorithms in our study. Since linear regression minimizes the sum of squares error, it also minimizes the Euclidean distance (the Euclidean

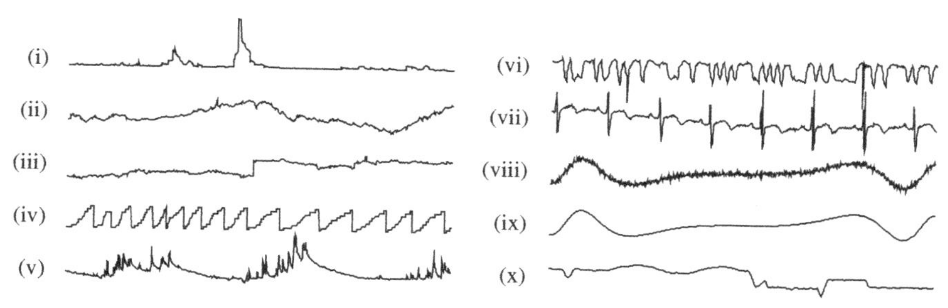


Fig. 3. The 10 datasets used in the experiments. (i) Radio Waves. (ii) Exchange Rates. (iii) Tickwise II. (iv) Tickwise I. (v) Water Level. (vi) Manufacturing. (vii) ECG. (viii) Noisy Sine Cubed. (ix) Sine Cube. (x) Space Shuttle.

distance is just the square root of the sum of squares). Euclidean distance, or some measure derived from it, is by far the most common metric used in data mining of time series [Agrawal *et al.* (1993), Agrawal *et al.* (1995), Chan and Fu (1999), Das *et al.* (1998), Keogh *et al.* (2000), Keogh and Pazzani (1999), Keogh and Pazzani (1998), Keogh and Smyth (1997), Qu *et al.* (1998), Wang and Wang (2000)]. The linear interpolation versions of the algorithms, by definition, will always have a greater sum of squares error.

We immediately encounter a problem when attempting to compare the algorithms. We cannot compare them for a fixed number of segments, since Sliding Windows does not allow one to specify the number of segments. Instead we give each of the algorithms a fixed `max_error` and measure the total error of the entire piecewise approximation.

The performance of the algorithms depends on the value of `max_error`. As `max_error` goes to zero all the algorithms have the same performance, since they would produce $n/2$ segments with no error. At the opposite end, as `max_error` becomes very large, the algorithms once again will all have the same performance, since they all simply approximate T with a single best-fit line. Instead, we must test the relative performance for some reasonable value of `max_error`, a value that achieves a good trade off between compression and fidelity. Because this "reasonable value" is subjective and dependent on the data mining application and the data itself, we did the following. We chose what we considered a "reasonable value" of `max_error` for each dataset, and then we bracketed it with 6 values separated by powers of two. The lowest of these values tends to produce an over-fragmented approximation, and the highest tends to produce a very coarse approximation. So in general, the performance in the mid-range of the 6 values should be considered most important. Figure 4 illustrates this idea.

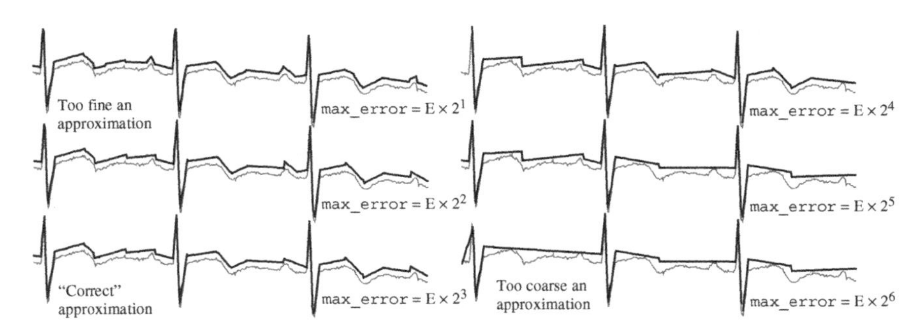


Fig. 4. We are most interested in comparing the segmentation algorithms at the setting of the user-defined threshold max_error that produces an intuitively correct level of approximation. Since this setting is subjective we chose a value for E, such that max_error $= \mathrm{E} \times 2^i$ ($i = 1$ to 6), brackets the range of reasonable approximations.

Since we are only interested in the relative performance of the algorithms, for each setting of `max_error` on each data set, we normalized the performance of the 3 algorithms by dividing by the error of the worst performing approach.

### 3.2. *Experimental Results*

The experimental results are summarized in Figure 5. The most obvious result is the generally poor quality of the Sliding Windows algorithm. With a few exceptions, it is the worse performing algorithm, usually by a large amount.

Comparing the results for Sine cubed and Noisy Sine supports our conjecture that the noisier a dataset, the less difference one can expect between algorithms. This suggests that one should exercise caution in attempting to generalize the performance of an algorithm that has only been demonstrated on a single noisy dataset [Qu *et al.* (1998), Wang and Wang (2000)]. Top-Down does occasionally beat Bottom-Up, but only by small amount. On the other hand Bottom-Up often significantly out performs Top-Down, especially on the ECG, Manufacturing and Water Level data sets.

## 4. A New Approach

Given the noted shortcomings of the major segmentation algorithms, we investigated alternative techniques. The main problem with the Sliding Windows algorithm is its inability to look ahead, lacking the global view of its offline (batch) counterparts. The Bottom-Up and the Top-Down

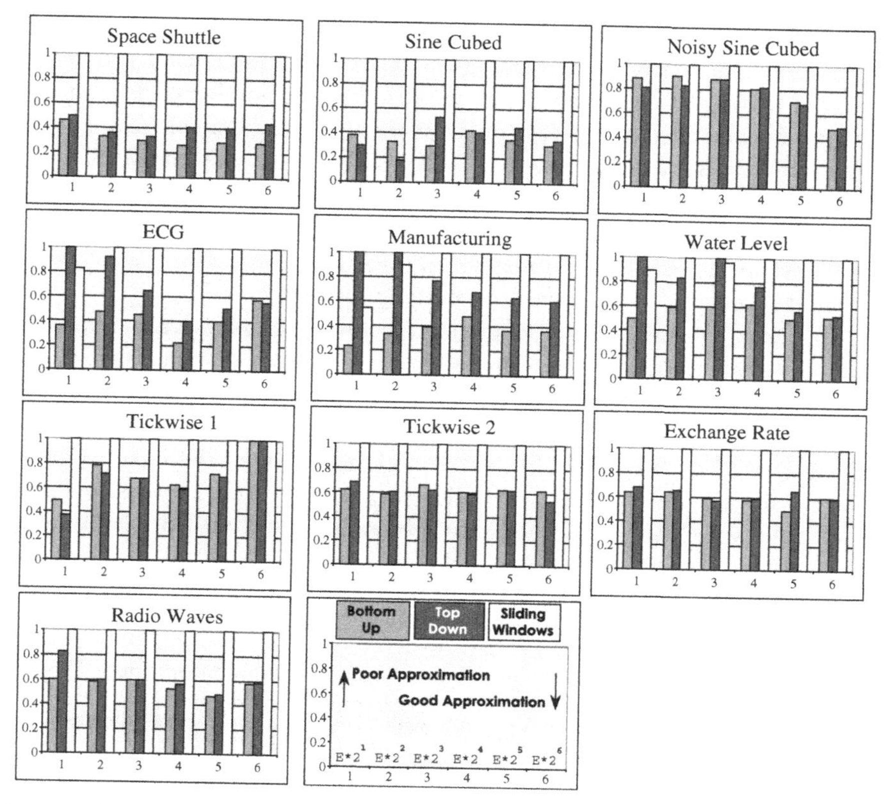


Fig. 5. A comparison of the three major times series segmentation algorithms, on ten diverse datasets, over a range in parameters. Each experimental result (i.e. a triplet of histogram bars) is normalized by dividing by the performance of the worst algorithm on that experiment.

approaches produce better results, but are offline and require the scanning of the entire data set. This is impractical or may even be unfeasible in a data-mining context, where the data are in the order of terabytes or arrive in continuous streams. We therefore introduce a novel approach in which we capture the online nature of Sliding Windows and yet retain the superiority of Bottom-Up. We call our new algorithm SWAB (Sliding Window and Bottom-up).

## 4.1. *The SWAB Segmentation Algorithm*

The SWAB algorithm keeps a buffer of size $w$. The buffer size should initially be chosen so that there is enough data to create about 5 or 6 segments.

Bottom-Up is applied to the data in the buffer and the leftmost segment is reported. The data corresponding to the reported segment is removed from the buffer and more datapoints are read in. The number of datapoints read in depends on the structure of the incoming data. This process is performed by the `Best_Line` function, which is basically just classic Sliding Windows. These points are incorporated into the buffer and Bottom-Up is applied again. This process of applying Bottom-Up to the buffer, reporting the leftmost segment, and reading in the next "best fit" subsequence is repeated as long as data arrives (potentially forever).

The intuition behind the algorithm is this. The `Best_Line` function finds data corresponding to a single segment using the (relatively poor) Sliding Windows and gives it to the buffer. As the data moves through the buffer the (relatively good) Bottom-Up algorithm is given a chance to refine the segmentation, because it has a "semi-global" view of the data. By the time the data is ejected from the buffer, the segmentation breakpoints are usually the same as the ones the batch version of Bottom-Up would have chosen. Table 6 shows the pseudo code for the algorithm.

Table 6. The SWAB (Sliding Window and Bottom-up) algorithm.

```
Algorithm Seg_TS = SWAB(max_error, seg_num) // seg_num is a small integer,
i.e. 5 or 6
read in w number of data points     // Enough to approximate
lower_bound = w / 2;     // seg_num of segments.
upper_bound = 2 * w;
while data at input
    T = Bottom_Up(w, max_error)     // Call the Bottom-Up algorithm.
    Seg_TS = CONCAT(SEG_TS, T(1));
    w = TAKEOUT(w, w');     // Deletes w' points in T(1) from w.
    if data at input     // Add w'' points from BEST_LINE() to w.
      w = CONCAT(w, BEST_LINE(max_error));
      {check upper and lower bound, adjust if necessary}
    else     // flush approximated segments from buffer.
      Seg_TS = CONCAT(SEG_TS, (T - T(1)))
    end;
end;
Function S = BEST_LINE(max_error)     // returns S points to approximate.
while error ≤ max_error     // next potential segment.
  read in one additional data point, d, into S
  S = CONCAT(S, d);
  error = approx_segment(S);
end while;
return S;
```

Using the buffer allows us to gain a "semi-global" view of the data set for Bottom-Up. However, it important to impose upper and lower bounds on the size of the window. A buffer that is allowed to grow arbitrarily large will revert our algorithm to pure Bottom-Up, but a small buffer will deteriorate it to Sliding Windows, allowing excessive fragmentation may occur. In our algorithm, we used an upper (and lower) bound of twice (and half) of the initial buffer.

Our algorithm can be seen as operating on a continuum between the two extremes of Sliding Windows and Bottom-Up. The surprising result (demonstrated below) is that by allowing the buffer to contain just 5 or 6 times the data normally contained by is a single segment, the algorithm produces essentially the same results as Bottom-Up, yet is able process a never-ending stream of data. Our new algorithm requires only a small, constant amount of memory, and the time complexity is a small constant factor worse than that of the standard Bottom-Up algorithm.

### 4.2. *Experimental Validation*

We repeated the experiments in Section 3, this time comparing the new algorithm with pure (batch) Bottom-Up and classic Sliding Windows. The result, summarized in Figure 6, is that the new algorithm produces results that are essentiality identical to Bottom-Up. The reader may be surprised that SWAB can sometimes be slightly better than Bottom-Up. The reason why this can occur is because SWAB is exploring a slight larger search space. Every segment in Bottom-Up must have an even number of data-points, since it was created by merging other segments that also had an even number of segments. The only possible exception is the rightmost segment, which can have an even number of segments if the original time series had an odd length. Since this happens multiple times for SWAB, it is effectively searching a slight larger search space.

## 5. Conclusions and Future Directions

We have seen the first extensive review and empirical comparison of time series segmentation algorithms from a data mining perspective. We have shown the most popular approach, Sliding Windows, generally produces very poor results, and that while the second most popular approach, Top-Down, can produce reasonable results, it does not scale well. In contrast, the least well known, Bottom-Up approach produces excellent results and scales linearly with the size of the dataset.

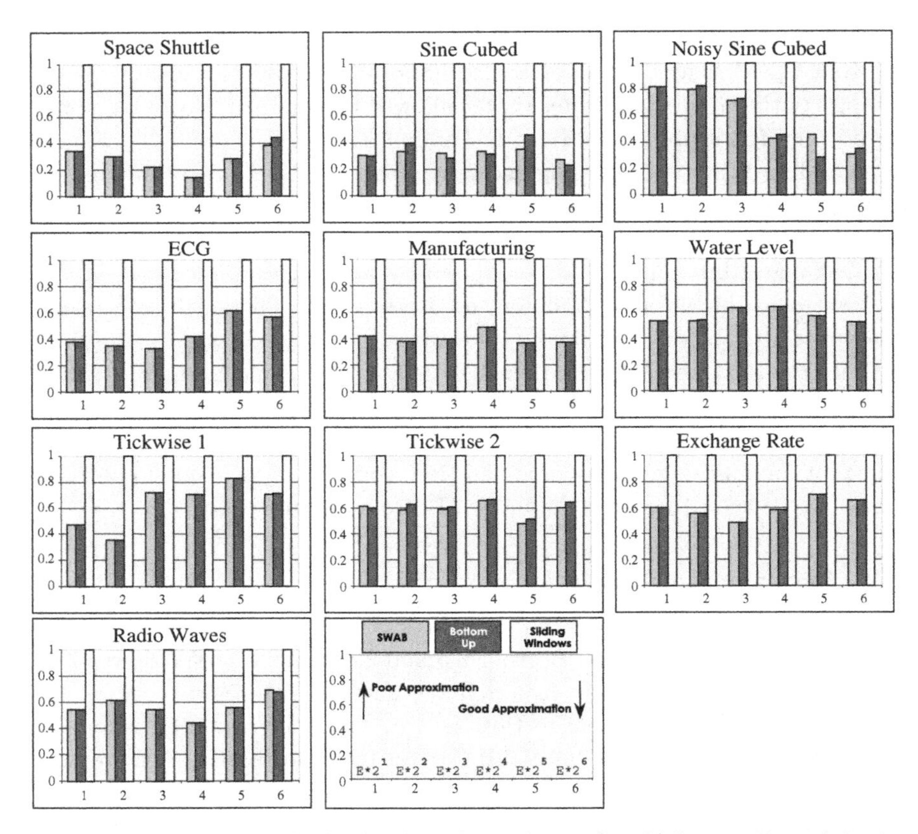


Fig. 6. A comparison of the SWAB algorithm with pure (batch) Bottom-Up and classic Sliding Windows, on ten diverse datasets, over a range in parameters. Each experimental result (i.e. a triplet of histogram bars) is normalized by dividing by the performance of the worst algorithm on that experiment.

In addition, we have introduced SWAB, a new online algorithm, which scales linearly with the size of the dataset, requires only constant space and produces high quality approximations of the data.

There are several directions in which this work could be expanded.

- The performance of Bottom-Up is particularly surprising given that it explores a smaller space of representations. Because the initialization phase of the algorithm begins with all line segments having length two, all merged segments will also have even lengths. In contrast the two other algorithms allow segments to have odd or even lengths. It would be interesting to see if removing this limitation of Bottom-Up can improve its performance further.

- For simplicity and brevity, we have assumed that the inner loop of the SWAB algorithm simply invokes the Bottom-Up algorithm each time. This clearly results in some computation redundancy. We believe we may be able to reuse calculations from previous invocations of Bottom-Up, thus achieving speedup.

**Reproducible Results Statement:** In the interests of competitive scientific inquiry, all datasets and code used in this work are freely available at the University of California Riverside, Time Series Data Mining Archive {www.cs.ucr.edu/~eamonn/TSDMA/index.html}.

## References

1. Agrawal, R., Faloutsos, C., and Swami, A. (1993). Efficient Similarity Search in Sequence Databases. *Proceedings of the 4th Conference on Foundations of Data Organization and Algorithms*, pp. 69–84.
2. Agrawal, R., Lin, K.I., Sawhney, H.S., and Shim, K. (1995). Fast Similarity Search in the Presence of Noise, Scaling, and Translation in Times-Series Databases. *Proceedings of 21th International Conference on Very Large Data Bases*, pp. 490–501.
3. Chan, K. and Fu, W. (1999). Efficient Time Series Matching by Wavelets. *Proceedings of the 15th IEEE International Conference on Data Engineering*, pp. 126–133.
4. Das, G., Lin, K. Mannila, H., Renganathan, G., and Smyth, P. (1998). Rule Discovery from Time Series. *Proceedings of the 3rd International Conference of Knowledge Discovery and Data Mining*, pp. 16–22.
5. Douglas, D.H. and Peucker, T.K. (1973). Algorithms for the Reduction of the Number of Points Required to Represent a Digitized Line or its Caricature. *Canadian Cartographer*, **10**(2) December, pp. 112–122.
6. Duda, R.O. and Hart, P.E. (1973). Pattern Classification and Scene Analysis. Wiley, New York.
7. Ge, X. and Smyth P. (2001). Segmental Semi-Markov Models for Endpoint Detection in Plasma Etching. *IEEE Transactions on Semiconductor Engineering.*
8. Heckbert, P.S. and Garland, M. (1997). Survey of Polygonal Surface Simplification Algorithms, Multiresolution Surface Modeling Course. *Proceedings of the 24th International Conference on Computer Graphics and Interactive Techniques.*
9. Hunter, J. and McIntosh, N. (1999). Knowledge-Based Event Detection in Complex Time Series Data. *Artificial Intelligence in Medicine*, Springer, pp. 271–280.

10. Ishijima, M.. *et al.* (1983). Scan-Along Polygonal Approximation for Data Compression of Electrocardiograms. *IEEE Transactions on Biomedical Engineering (BME)*, **30**(11), 723–729.
11. Koski, A., Juhola, M., and Meriste, M. (1995). Syntactic Recognition of ECG Signals By Attributed Finite Automata. *Pattern Recognition*, **28**(12), 1927–1940.
12. Keogh, E., Chakrabarti, K., Pazzani, M., and Mehrotra, S. (2000). Dimensionality Reduction for Fast Similarity Search in Large Time Series Databases. *Journal of Knowledge and Information Systems*, **3**(3), 263–286.
13. Keogh, E. and Pazzani, M. (1998). An Enhanced Representation of Time Series which Allows Fast and Accurate Classification, Clustering and Relevance Feedback. *Proceedings of the 4th International Conference of Knowledge Discovery and Data Mining*, AAAI Press, pp. 239–241.
14. Keogh, E. and Pazzani, M. (1999). Relevance Feedback Retrieval of Time Series Data. *Proceedings of the 22th Annual International ACM-SIGIR Conference on Research and Development in Information Retrieval*, pp. 183–190.
15. Keogh, E. and Smyth, P. (1997). A Probabilistic Approach to Fast Pattern Matching in Time Series Databases. *Proceedings of the 3rd International Conference of Knowledge Discovery and Data Mining*, pp. 24–20.
16. Last, M., Klein, Y., and Kandel, A. (2001). Knowledge Discovery in Time Series Databases. *IEEE Transactions on Systems, Man, and Cybernetics*, **31B**(1), 160–169.
17. Lavrenko, V., Schmill, M., Lawrie, D., Ogilvie, P., Jensen, D., and Allan, J. (2000). Mining of Concurrent Text and Time Series. *Proceedings of the 6th International Conference on Knowledge Discovery and Data Mining*, 37–44.
18. Li, C,. Yu, P., and Castelli, V. (1998). MALM: A Framework for Mining Sequence Database at Multiple Abstraction Levels. *Proceedings of the 9th International Conference on Information and Knowledge Management*, pp. 267–272.
19. McKee, J.J, Evans, N.E, and Owens, F.J (1994). Efficient Implementation of the Fan/SAPA-2 Algorithm Using Fixed Point Arithmetic. *Automedica*, **16**, 109–117.
20. Osaki, R., Shimada, M., and Uehara, K. (1999). Extraction of Primitive Motion for Human Motion Recognition. *Proceedings of the 2nd International Conference on Discovery Science*, pp. 351–352.
21. Park, S., Kim, S.W, and Chu, W.W (2001). Segment-Based Approach for Subsequence Searches in Sequence Databases. *Proceedings of the 16th ACM Symposium on Applied Computing*, pp. 248–252.
22. Park, S., Lee, D., and Chu, W.W (1999). Fast Retrieval of Similar Subsequences in Long Sequence Databases. *Proceedings of the 3rd IEEE Knowledge and Data Engineering Exchange Workshop.*
23. Pavlidis, T. (1976). Waveform Segmentation Through Functional Approximation. *IEEE Transactions on Computers*, pp. 689–697.
24. Perng, C., Wang, H., Zhang, S., and Parker, S. (2000). Landmarks: A New Model for Similarity-Based Pattern Querying in Time Series Databases. *Proceedings of 16th International Conference on Data Engineering*, pp. 33–45.

25. Qu, Y., Wang, C., and Wang, S. (1998). Supporting Fast Search in Time Series for Movement Patterns in Multiples Scales, *Proceedings of the 7th International Conference on Information and Knowledge Management*, pp. 251–258.
26. Ramer, U. (1972). An Iterative Procedure for the Polygonal Approximation of Planar Curves. *Computer Graphics and Image Processing*, **1**, 244–256.
27. Shatkay, H. (1995). Approximate Queries and Representations for Large Data Sequences. *Technical Report cs-95-03*, Department of Computer Science, Brown University.
28. Shatkay, H. and Zdonik, S. (1996). Approximate Queries and Representations for Large Data Sequences. *Proceedings of the 12th IEEE International Conference on Data Engineering*, pp. 546–553.
29. Sugiura, N. and Ogden, R.T (1994). Testing Change-Points with Linear Trend. *Communications in Statistics B: Simulation and Computation*, **23**, 287–322.
30. Vullings, H.J L.M., Verhaegen, M.H.G., and Verbruggen H.B. (1997). ECG Segmentation Using Time-Warping. *Proceedings of the 2nd International Symposium on Intelligent Data Analysis*, pp. 275–286.
31. Wang, C. and Wang, S. (2000). Supporting Content-Based Searches on Time Series Via Approximation. *Proceedings of the 12th International Conference on Scientific and Statistical Database Management*, pp. 69–81.

## CHAPTER 2

# A SURVEY OF RECENT METHODS FOR EFFICIENT RETRIEVAL OF SIMILAR TIME SEQUENCES

Magnus Lie Hetland
*Norwegian University of Science and Technology*
*Sem Sælands vei 7–9*
*NO-7491 Trondheim, Norway*
E-mail: magnus@hetland.org

Time sequences occur in many applications, ranging from science and technology to business and entertainment. In many of these applications, searching through large, unstructured databases based on sample sequences is often desirable. Such similarity-based retrieval has attracted a great deal of attention in recent years. Although several different approaches have appeared, most are based on the common premise of dimensionality reduction and spatial access methods. This chapter gives an overview of recent research and shows how the methods fit into a general context of signature extraction.

*Keywords*: Information retrieval; sequence databases; similarity search; spatial indexing; time sequences.

## 1. Introduction

Time sequences arise in many applications—any applications that involve storing sensor inputs, or sampling a value that changes over time. A problem which has received an increasing amount of attention lately is the problem of *similarity retrieval* in databases of time sequences, so-called "query by example." Some uses of this are [Agrawal *et al.* (1993)]:

- Identifying companies with similar patterns of growth.
- Determining products with similar selling patterns.
- Discovering stocks with similar movement in stock prices.

- Finding out whether a musical score is similar to one of a set of copyrighted scores.
- Finding portions of seismic waves that are not similar to spot geological irregularities.

Applications range from medicine, through economy, to scientific disciplines such as meteorology and astrophysics [Faloutsos *et al.* (1994), Yi and Faloutsos (2000)].

The running times of simple algorithms for comparing time sequences are generally polynomial in the length of both sequences, typically linear or quadratic. To find the correct offset of a query in a large database, a naive *sequential scan* will require a number of such comparisons that is linear in the length of the database. This means that, given a query of length $m$ and a database of length $n$, the search will have a time complexity of $O(nm)$, or even $O(nm^2)$ or worse. For large databases this is clearly unacceptable.

Many methods are known for performing this sort of query in the domain of strings over finite alphabets, but with time sequences there are a few extra issues to deal with:

- The range of values is not generally finite, or even discrete.
- The sampling rate may not be constant.
- The presence of noise in various forms makes it necessary to support very flexible similarity measures.

This chapter describes some of the recent advances that have been made in this field; methods that allow for indexing of time sequences using flexible similarity measures that are invariant under a wide range of transformations and error sources.

The chapter is structured as follows: Section 2 gives a more formal presentation of the problem of similarity-based retrieval and the so-called dimensionality curse; Section 3 describes the general approach of signature based retrieval, or shrink and search, as well as three specific methods using this approach; Section 4 shows some other approaches, while Section 5 concludes the chapter. Finally, Appendix gives an overview of some basic distance measures.[1]

[1]The term "distance" is used loosely in this paper. A distance measure is simply the inverse of a similarity measure and is not required to obey the metric axioms.

### 1.1. *Terminology and Notation*

A time sequence $\vec{x} = \langle x_1 = (v_1, t_1), \ldots, x_n = (v_n, t_n)\rangle$ is an ordered collection of elements $x_i$, each consisting of a value $v_i$ and a timestamp $t_i$. Abusing the notation slightly, the value of $x_i$ may be referred to as $x_i$.

For some retrieval methods, the values may be taken from a finite class of values [Mannila and Ronkainen (1997)], or may have more than one dimension [Lee *et al.* (2000)], but it is generally assumed that the values are real numbers. This assumption is a requirement for most of the methods described in this chapter.

The only requirement of the timestamps is that they be non-decreasing (or, in some applications, strictly increasing) with respect to the sequence indices:

$$t_i \leq t_j \Leftrightarrow i \leq j. \tag{1}$$

In some methods, an additional assumption is that the elements are *equi-spaced*: for every two consecutive elements $x_i$ and $x_{i+1}$ we have

$$t_{i+1} - t_i = \Delta, \tag{2}$$

where $\Delta$ (the *sampling rate* of $\vec{x}$) is a (positive) constant. If the actual sampling rate is not important, $\Delta$ may be normalized to 1, and $t_1$ to 0. It is also possible to resample the sequence to make the elements equi-spaced, when required.

The *length* of a time sequence $\vec{x}$ is its cardinality, written as $|\vec{x}|$. The contiguous subsequence of $\vec{x}$ containing elements $x_i$ to $x_j$ (inclusive) is written $x_{i:j}$. A *signature* of a sequence $\vec{x}$ is some structure that somehow represents $\vec{x}$, yet is simpler than $\vec{x}$. In the context of this chapter, such a signature will always be a vector of fixed size $k$. (For a more thorough discussion of signatures, see Section 3.) Such a signature is written $\tilde{x}$. For a summary of the notation, see Table 1.

Table 1. Notation.

| | |
|---|---|
| $\vec{x}$ | A sequence |
| $\tilde{x}$ | A signature of $\vec{x}$ |
| $x_i$ | Element number $i$ of $\vec{x}$ |
| $x_{i:j}$ | Elements $i$ to $j$ (inclusive) of $\vec{x}$ |
| $\lvert\vec{x}\rvert$ | The length of $\vec{x}$ |

## 2. The Problem

The problem of retrieving similar time sequences may be stated as follows: Given a sequence $\vec{q}$, a set of time sequences $X$, a (non-negative) distance measure $d$, and a *tolerance threshold* $\varepsilon$, find the set $R$ of sequences closer to $\vec{q}$ than $\varepsilon$, or, more precisely:

$$R = \{\vec{x} \in X | d(\vec{q}, \vec{x}) \leq \varepsilon\}. \tag{3}$$

Alternatively, one might wish to find the $k$ nearest neighbours of $\vec{q}$, which amounts to setting $\varepsilon$ so that $|R| = k$. The parameter $\varepsilon$ is typically supplied by the user, while the distance function $d$ is domain-dependent. Several distance measures will be described rather informally in this chapter. For more formal definitions, see Appendix.

Figure 1 illustrates the problem for Euclidean distance in two dimensions. In this example, the vector $\vec{x}$ will be included in the result set $R$, while $\vec{y}$ will not.

A useful variation of the problem is to find a set of *subsequences* of the sequences in $X$. This, in the basic case, requires comparing $\vec{q}$ not only to all elements of $X$, but to all possible subsequences.[2]

If a method retrieves a subset $S$ of $R$, the wrongly dismissed sequences in $R - S$ are called *false dismissals*. Conversely, if $S$ is a superset of $R$, the sequences in $S - R$ are called *false alarms*.

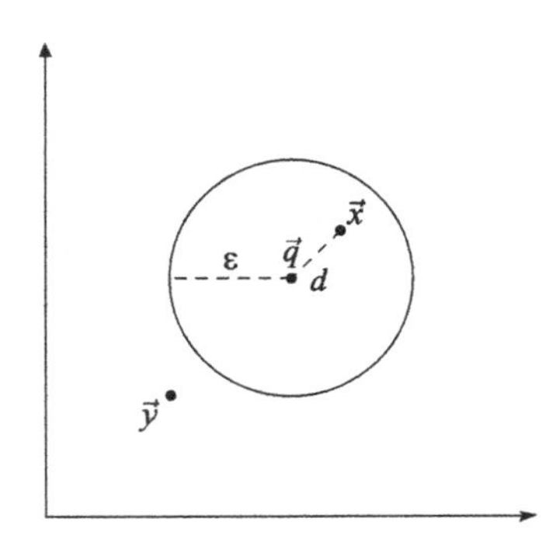


Fig. 1. Similarity retrieval.

[2]Except in the description of LCS in Appendix, *subsequence* means *contiguous subsequence*, or *segment*.

### 2.1. *Robust Distance Measures*

The choice of distance measure is highly domain dependent, and in some cases a simple $L_p$ norm such as Euclidean distance may be sufficient. However, in many cases, this may be too brittle [Keogh and Pazzani (1999b)] since it does not tolerate such transformations as scaling, warping, or translation along either axis. Many of the newer retrieval methods focus on using more robust distance measures, which are invariant under such transformations as *time warping* (see Appendix for details) without loss of performance.

### 2.2. *Good Indexing Methods*

Faloutsos *et al.* (1994) list the following desirable properties for an indexing method:

(i) It should be faster than a sequential scan.
(ii) It should incur little space overhead.
(iii) It should allow queries of various length.
(iv) It should allow insertions and deletions without rebuilding the index.
(v) It should be correct: No false dismissals must occur.

To achieve high performance, the number of false alarms should also be low. Keogh *et al.* (2001b) add the following criteria to the list above:

(vi) It should be possible to build the index in reasonable time.
(vii) The index should preferably be able to handle more than one distance measure.

### 2.3. *Spatial Indices and the Dimensionality Curse*

The general problem of similarity based retrieval is well known in the field of information retrieval, and many indexing methods exist to process queries efficiently [Baeza-Yates and Ribeiro-Neto (1999)]. However, certain properties of time sequences make the standard methods unsuitable. The fact that the value ranges of the sequences usually are continuous, and that the elements may not be equi-spaced, makes it difficult to use standard text-indexing techniques such as suffix-trees. One of the most promising techniques is multidimensional indexing ($R$-trees [Guttman (1984)], for instance), in which the objects in question are multidimensional vectors, and similar objects can be retrieved in sublinear time. One requirement of such spatial access methods is that the distance measure must be monotonic in all dimensions, usually satisfied through the somewhat stricter requirement of the triangle inequality ($d(\vec{x}, \vec{z}) \leq d(\vec{x}, \vec{y}) + d(\vec{y}, \vec{z})$).

One important problem that occurs when trying to index sequences with spatial access methods is the so-called *dimensionality curse*: Spatial indices typically work only when the number of dimensions is low [Chakrabarti and Mehrotra (1999)]. This makes it unfeasible to code the entire sequence directly as a vector in an indexed space.

The general solution to this problem is *dimensionality reduction*: to condense the original sequences into *signatures* in a *signature space* of low dimensionality, in a manner which, to some extent, preserves the distances between them. One can then index the signature space.

## 3. Signature Based Similarity Search

A time sequence $\vec{x}$ of length $n$ can be considered a vector or point in an $n$-dimensional space. Techniques exist (spatial access methods, such as the $R$-tree and variants [Chakrabarti and Mehrotra (1999), Wang and Perng (2001), Sellis *et al.* (1987)] for indexing such data. The problem is that the performance of such methods degrades considerably even for relatively low dimensionalities [Chakrabarti and Mehrotra (1999)]; the number of dimensions that can be handled is usually several orders of magnitude lower than the number of data points in a typical time sequence.

A general solution described by Faloutsos *et al.* (1994; 1997) is to extract a low-dimensional *signature* from each sequence, and to index the signature space. This *shrink* and *search* approach is illustrated in Figure 2.

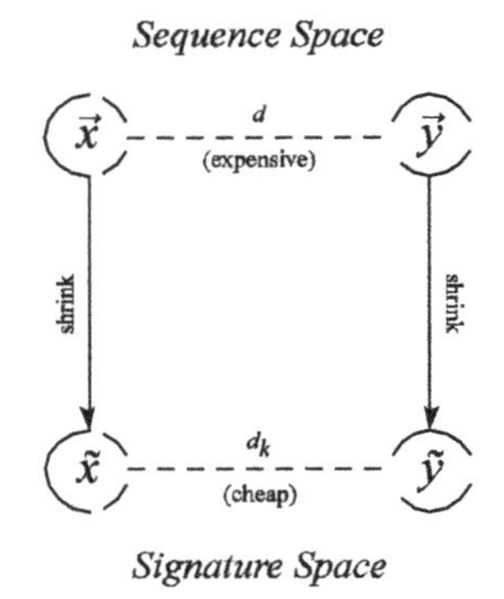


Fig. 2. The signature based approach.

An important result given by Faloutsos *et al.* (1994) is the proof that in order to guarantee completeness (no false dismissals), the distance function used in the signature space must underestimate the true distance measure, or:

$$d_k(\tilde{x}, \tilde{y}) \le d(\vec{x}, \vec{y}). \tag{4}$$

This requirement is called the *bounding lemma*. Assuming that (1.4) holds, an intuitive way of stating the resulting situation is: "if two signatures are far apart, we know the corresponding [sequences] must also be far apart" [Faloutsos *et al.* (1997)]. This, of course, means that there will be no false dismissals. To minimise the number of false alarms, we want $d_k$ to approximate $d$ as closely as possible. The bounding lemma is illustrated in Figure 3.

This general method of dimensionality reduction may be summed up as follows [Keogh *et al.* (2001b)]:

1. Establish a distance measure $d$ from a domain expert.
2. Design a dimensionality reduction technique to produce signatures of length $k$, where $k$ can be efficiently handled by a standard spatial access method.
3. Produce a distance measure $d_k$ over the $k$-dimensional signature space, and prove that it obeys the bounding condition (4).

In some applications, the requirement in (4) is relaxed, allowing for a small number of false dismissals in exchange for increased performance. Such methods are called *approximate*.

The dimensionality reduction may in itself be used to speed up the sequential scan, and some methods (such as the piecewise linear approximation of Keogh *et al.*, which is described in Section 4.2) rely only on this, without using any index structure.

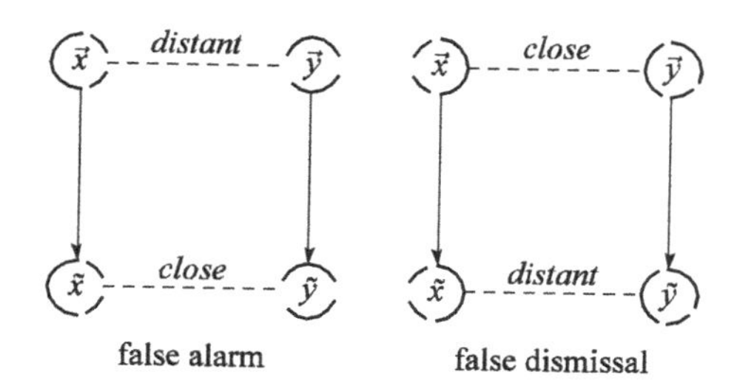


Fig. 3. An intuitive view of the bounding lemma.

Methods exist for finding signatures of arbitrary objects, given the distances between them [Faloutsos and Lin (1995), Wang *et al.* (1999)], but in the following I will concentrate on methods that exploit the structure of the time series to achieve good approximations.

### 3.1. *A Simple Example*

As an example of the signature based scheme, consider the two sequences shown in Figure 4.

The sequences, $\vec{x}$ and $\vec{y}$, are compared using the $L_1$ measure (Manhattan distance), which is simply the sum of the absolute distances between each aligning pair of values. A simple signature in this scheme is the prefix of length 2, as indicated by the shaded area in the figure. As shown in Figure 5, these signatures may be interpreted as points in a two-dimensional plane, which can be indexed with some standard spatial indexing method. It is also clear that the signature distance will underestimate the real distance between the sequences, since the remaining summands of the real distance must all be positive.

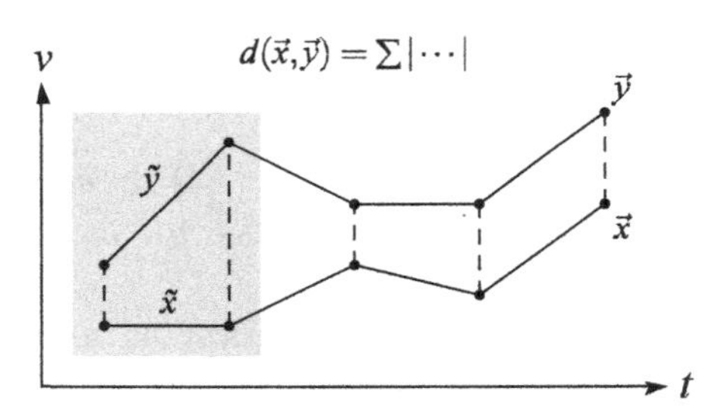


Fig. 4. Comparing two sequences.

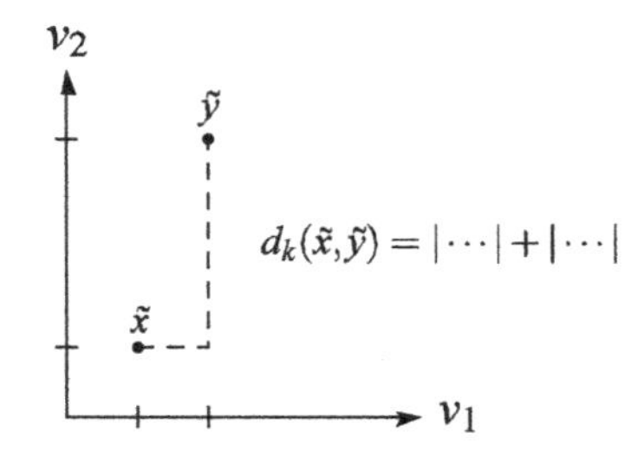


Fig. 5. A simple signature distance.

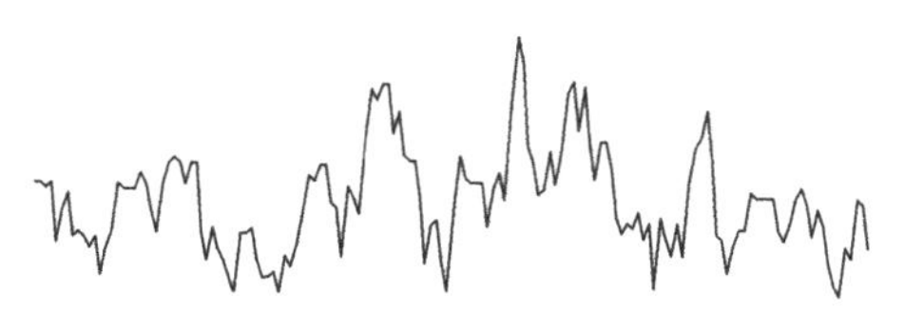

Fig. 6. An example time sequence.

Although correct, this simple signature extraction technique is not particularly precise. The signature extraction methods introduced in the following sections take into account more information about the full sequence shape, and therefore lead to fewer false alarms.

Figure 6 shows a time series containing measurements of atmospheric pressure. In the following three sections, the methods described will be applied to this sequence, and the resulting simplified sequence (reconstructed from the extracted signature) will be shown superimposed on the original.

### 3.2. *Spectral Signatures*

Some of the methods presented in this section are not very recent, but introduce some of the main concepts used by newer approaches.

Agrawal *et al.* (1993) introduce a method called the $F$-index in which a signature is extracted from the frequency domain of a sequence. Underlying their approach are two key observations:

- Most real-world time sequences can be faithfully represented by their strongest Fourier coefficients.
- Euclidean distance is preserved in the frequency domain (Parseval's Theorem [Shatkay (1995)]).

Based on this, they suggest performing the Discrete Fourier Transform on each sequence, and using a vector consisting of the sequence's $k$ first amplitude coefficients as its signature. Euclidean distance in the signature space will then underestimate the real Euclidean distance between the sequences, as required.

Figure 7 shows an approximated time sequence, reconstructed from a signature consisting of the original sequence's ten first Fourier components.

This basic method allows only for whole-sequence matching. In 1994, Faloutsos *et al.* introduce the $ST$-index, an improvement on the $F$-index

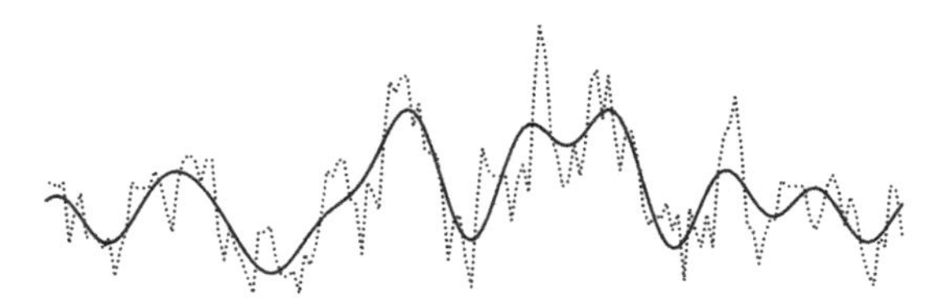

Fig. 7. A sequence reconstructed from a spectral signature.

that makes subsequence matching possible. The main steps of the approach are as follows:

1. For each position in the database, extract a window of length $w$, and create a spectral signature (a *point*) for it.

   Each point will be close to the previous, because the contents of the sliding window change slowly. The points for one sequence will therefore constitute a *trail* in signature space.

2. Partition the trails into suitable (multidimensional) Minimal Bounding Rectangles (MBRs), according to some heuristic.
3. Store the MBRs in a spatial index structure.

To search for subsequences similar to a query $\vec{q}$ of length $w$, simply look up all MBRs that intersect a hypersphere with radius $\varepsilon$ around the signature point $\tilde{q}$. This is guaranteed not to produce any false dismissals, because if a point is within a radius of $\varepsilon$ of $\tilde{q}$, it cannot possibly be contained in an MBR that does not intersect the hypersphere.

To search for sequences longer than $w$, split the query into $w$-length segments, search for each of them, and intersect the result sets. Because a sequence in the result set R cannot be closer to the full query sequence than it is to any one of the window signatures, it has to be close to all of them, that is, contained in all the result sets.

These two papers [Agrawal *et al.* (1993) and Faloutsos *et al.* (1994)] are seminal; several newer approaches are based on them. For example, Rafiei and Mendelzon (1997) show how the method can be made more robust by allowing various transformations in the comparison, and Chan and Fu (1999) show how the Discrete Wavelet Transform (DWT) can be used instead of the Discrete Fourier Transform (DFT), and that the DWT method is empirically superior. See Wu *et al.* (2000) for a comparison between similarity search based on DFT and DWT.

### 3.3. *Piecewise Constant Approximation*

An approach independently introduced by Yi and Faloutsos (2000) and Keogh *et al.* (2001b), Keogh and Pazzani (2000) is to divide each sequence into $k$ segments of equal length, and to use the average value of each segment as a coordinate of a $k$-dimensional signature vector. Keogh *et al.* call the method *Piecewise Constant Approximation*, or PCA. This deceptively simple dimensionality reduction technique has several advantages [Keogh *et al.* (2001b)]: The transform itself is faster than most other transforms, it is easy to understand and implement, it supports more flexible distance measures than Euclidean distance, and the index can be built in linear time.

Figure 8 shows an approximated time sequence, reconstructed from a ten-dimensional PCA signature.

Yi and Faloutsos (2000) also show that this signature can be used with arbitrary $L_p$ norms without changing the index structure, which is something no previous method [such as Agrawal *et al.* (1993; 1995), Faloutsos *et al.* (1994; 1997), Rafiei and Mendelzon (1997), or Yi *et al.* (1998)] could accomplish. This means that the distance measure may be specified by the user. Preprocessing to make the index more robust in the face of such transformations as *offset translation*, *amplitude scaling*, and *time scaling* can also be performed.

Keogh *et al.* demonstrate that the representation can also be used with the so-called *weighted Euclidean distance*, where each part of the sequence has a different weight.

Empirically, the PCA methods seem promising: Yi and Faloutsos demonstrate up to a ten times speedup over methods based on the discrete wavelet transform. Keogh *et al.* do not achieve similar speedups, but point to the fact that the structure allows for more flexible distance measures than many of the competing methods.

Keogh *et al.* (2001a) later propose an improved version of the PCA, the so-called *Adaptive Piecewise Constant Approximation*, or APCA. This is

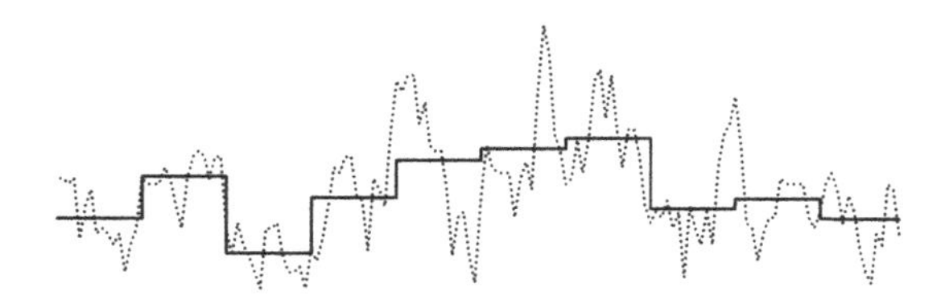

Fig. 8. A sequence reconstructed from a PCA signature.

similar to the PCA, except that the segments need not be of equal length. Thus regions with great fluctuations may be represented with several short segments, while reasonably featureless regions may be represented with fewer, long segments. The main contribution of this representation is that it is a more effective compression than the PCA, while still representing the original faithfully.

Two distance measures are developed for the APCA, one which is guaranteed to underestimate Euclidean distance, and one which can be calculated more efficiently, but which may generate some false dismissals. It is also shown that this technique, like the PCA, can handle arbitrary $L_p$ norms. The empirical data suggest that the APCA outperforms both methods based on the discrete Fourier transform, and methods based on the discrete wavelet transform with a speedup of one to two orders of magnitude.

In a recent paper, Keogh (2002) develops a distance measure that is a lower bound for dynamic time warping, and uses the PCA approach to index it. The distance measure is based on the assumption that the allowed warping is restricted, which is often the case in real applications. Under this assumption, Keogh constructs two warped versions of the sequence to be indexed: An upper and a lower limit. The PCA signatures of these limits are then extracted, and together with Keogh's distance measure form an exact index (one with no false dismissals) with high precision. Keogh performs extensive empirical experiments, and his method clearly outperforms any other existing method for indexing time warping.

### 3.4. *Landmark Methods*

In 1997, Keogh and Smyth introduce a probabilistic method for sequence retrieval, where the features extracted are characteristic parts of the sequence, so-called *feature shapes*. Keogh (1997) uses a similar *landmark based* technique. Both these methods also use the dimensionality reduction technique of piecewise linear approximation (see Section 4.2) as a preprocessing step. The methods are based on finding similar landmark features (or shapes) in the target sequences, ignoring shifting and scaling within given limits. The technique is shown to be significantly faster than sequential scanning (about an order of magnitude), which may be accounted for by the compression of the piecewise linear approximation. One of the contributions of the method is that it is one of the first that allows some longitudinal scaling.

A more recent paper by Perng *et al.* (2000) introduces a more general landmark model. In its most general form, the model allows any point of

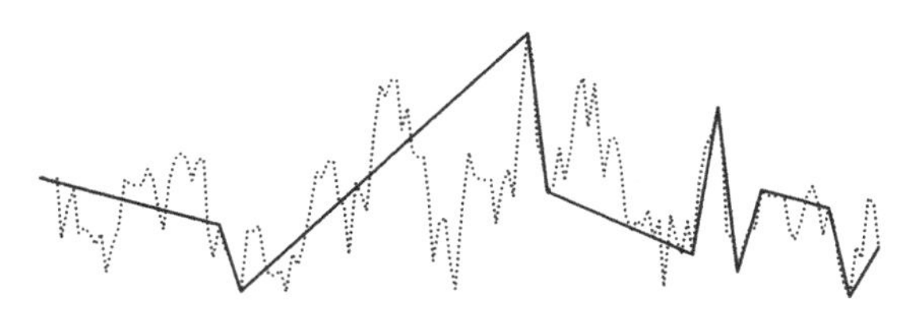

Fig. 9. A landmark approximation.

great importance to be identified as a landmark. The specific form used in the paper defines an $n$th order landmark of a one-dimensional function to be a point where the function's $n$th derivative is zero. Thus, first-order landmarks are extrema, second-order landmarks are inflection points, and so forth. A smothing technique is also introduced, which lets certain landmarks be overshadowed by others. For instance, local extrema representing small fluctuations may not be as important as a global maximum or minimum.

Figure 9 shows an approximated time sequence, reconstructed from a twelve-dimensional landmark signature.

One of the main contributions of Perng *et al.* (2000) is to show that for suitable selections of landmark features, the model is invariant with respect to the following transformations:

- Shifting
- Uniform amplitude scaling
- Uniform time scaling
- Non-uniform time scaling (time warping)
- Non-uniform amplitude scaling

It is also possible to allow for several of these transformations at once, by using the intersection of the features allowed for each of them. This makes the method quite flexible and robust, although as the number of transformations allowed increases, the number of features will decrease; consequently, the index will be less precise.

A particularly simple landmark based method (which can be seen as a special case of the general landmark method) is introduced by Kim *et al.* (2001). They show that by extracting the minimum, maximum, and the first and last elements of a sequence, one gets a (rather crude) signature that is invariant to time warping. However, since time warping distance does not obey the triangle inequality [Yi *et al.* (1998)], it cannot be used directly. This problem is solved by developing a new distance measure that underestimates the time warping distance while simultaneously satisfying the triangle inequality. Note that this method does not achieve results comparable to those of Keogh (2002).

## 4. Other Approaches

Not all recent methods rely on spatial access methods. This section contains a sampling of other approaches.

### 4.1. *Using Suffix Trees to Avoid Redundant Computation*

Baeza-Yates and Gonnet (1999) and Park *et al.* (2000) independently introduce the idea of using suffix trees [Gusfield (1997)] to avoid duplicate calculations when using dynamic programming to compare a query sequence with other sequences in a database. Baeza-Yates and Gonnet use *edit distance* (see Appendix for details), while Park *et al.* use time warping.

The basic idea of the approach is as follows: When comparing two sequences $\vec{x}$ and $\vec{y}$ with dynamic programming, a subtask will be to compare their prefixes $x_{1:i}$ and $y_{1:j}$. If two other sequences are compared that have identical prefixes to these (for instance, the query and another sequence from the database), the same calculations will have to be performed again. If a sequential search for subsequence matches is performed, the cost may easily become prohibitive.

To avoid this, all the sequences in the database are indexed with a suffix tree. A suffix tree stores all the suffixes of a sequence, with identical prefixes stored only once. By performing a depth-first traversal of the suffix tree one can access every suffix (which is equivalent to each possible subsequence position) and backtrack to reuse the calculations that have already been performed for the prefix that the current and the next candidate subsequence share.

Baeza-Yates and Gonnet assume that the sequences are strings over a finite alphabet; Park *et al.* avoid this assumption by classifying each sequence element into one of a finite set of categories. Both methods achieve subquadratic running times.

### 4.2. *Data Reduction through Piecewise Linear Approximation*

Keogh *et al.* have introduced a dimensionality reduction technique using piecewise linear approximation of the original sequence data [Keogh (1997), Keogh and Pazzani (1998), Keogh and Pazzani (1999a), Keogh and Pazzani (1999b), Keogh and Smyth (1997)]. This reduces the number of data points by a compression factor typically in the range from 10 to 600 for real data [Keogh (1997)], outperforming methods based on the Discrete Fourier Transform by one to three orders of magnitude [Keogh and Pazzani (1999b)]. This approximation is shown to be valid under several distance measures, including dynamic time warping distance [Keogh and Pazzani (1999b)]. An enhanced representation is introduced in [Keogh and Pazzani (1998)], where every line segment in the approximation is augmented with a weight representing its relative importance; for instance, a combined sequence may be constructed representing a class of sequences, and some line segments may be more representative of the class than others.

### 4.3. *Search Space Pruning through Subsequence Hashing*

Keogh and Pazzani (1999a) describe an indexing method based on hashing, in addition to the piecewise linear approximation. An equi-spaced template grid window is moved across the sequence, and for each position a hash key is generated to decide into which *bin* the corresponding subsequence is put. The hash key is simply a binary string, where 1 means that the sequence is predominantly increasing in the corresponding part of the template grid, while 0 means that it is decreasing. These bin keys may then be used during a search, to prune away entire bins without examining their contents. To get more benefit from the bin pruning, the bins are arranged in a *best-first* order.

## 5. Conclusion

This chapter has sought to give an overview of recent advances in the field of similarity based retrieval in time sequence databases. First, the problem of similarity search and the desired properties of robust distance measures and good indexing methods were outlined. Then, the general approach of signature based similarity search was described. Following the general description, three specific signature extraction approaches were discussed: Spectral signatures, based on Fourier components (or wavelet components); piecewise constant approximation, and the related method adaptive piecewise constant approximation; and landmark methods, based on the extraction of significant points in a sequence. Finally, some methods that are not based on signature extraction were mentioned.

Although the field of time sequence indexing has received much attention and is now a relatively mature field [Keogh *et al.* (2002)] there are still areas where further research might be warranted. Two such areas are (1) thorough empirical comparisons and (2) applications in data mining.

The published methods have undergone thorough empirical tests that evaluate their performance (usually by comparing them to sequential scan, and, in some cases, to the basic spectral signature methods), but comparing the results is not a trivial task—in most cases it might not even be very meaningful, since variations in performance may be due to implementation details, available hardware, and several other factors that may not be inherent in the indexing methods themselves. Implementing several of the most promising methods and testing them on real world problems (under similar conditions) might lead to new insights, not only about their relative performances in general, but also about which methods are best suited for which problems. Although some comparisons have been made [such as in Wu *et al.* (2000) and, in the more general context of spatial similarity search, in Weber *et al.* (1998)], little research seems to have been published on this topic.

Data mining in time series databases is a relatively new field [Keogh *et al.* (2002)]. Most current mining methods are based on a full, linear scan of the sequence data. While this may seem unavoidable, constructing an index of the data could make it possible to perform this full data traversal only once, and later perform several data mining passes that only use the index to perform their work. It has been argued that data mining should be interactive [Das *et al.* (1998)], in which case such techniques could prove useful. Some publications can be found about using time sequence indexing for data mining purposes [such as Keogh *et al.* (2002), where a method is presented for mining patterns using a suffix tree index] but there is still a potential for combining existing sequence mining techniques with existing methods for similarity-based retrieval.

## Appendix Distance Measures

Faloutsos *et al.* (1997) describe a general framework for sequence distance measures [a similar framework can be found in Jagadish *et al.* (1995)]. They show that many common distance measures can be expressed in the following form:

$$d(\vec{x}, \vec{y}) = \min \begin{cases} \min_{T_1, T_2 \in T} \{c(T_1) + c(T_2) + d(T_1(\vec{x}), T_2(\vec{y}))\} \\ d_0(\vec{x}, \vec{y}). \end{cases} \tag{5}$$

$T$ is a set of allowable transformations, $c(T_i)$ is the *cost* of performing the transformation $T_i$, $T_i(\vec{x})$ is the sequence resulting from performing the transformation $T_i$ on $\vec{x}$, and $d_0$ is a so-called *base distance*, typically calculated in linear time. For instance, $L_p$ norms (such as Manhattan distance and Euclidean distance) results when $T = \emptyset$ and

$$d_0(\vec{x}\ \vec{y}) = L_p = \sqrt[p]{\sum\nolimits_{i=1}^{l} |x_i - y_i|^p} \tag{6}$$

where $|\vec{x}| = |\vec{y}| = l$.

Editing distance (or Levenshtein distance) is the weight of the minimum sequence of editing operations needed to transform one sequence into another [Sankoff and Kruskal (1999)]. It is usually defined on strings (or equi-spaced time sequences), but Mannila and Ronkainen (1997) show how to generalise this measure to general (non equi-spaced) time sequences. In the framework given above, editing distance may be defined as:

$$d_{ed}(\vec{x}, \vec{y}) = \min \begin{cases} c(\text{del}(x_1)) + d_{ed}(x_{2:m}, \vec{y}) \\ c(\text{del}(y_1)) + d_{ed}(\vec{x}, y_{2:n}) \\ c(\text{sub}(x_1\, y_1)) + d_{ed}(x_{2:m}, y_{2:n}) \end{cases} \tag{7}$$

where $m = |\vec{x}|$, $n = |\vec{y}|$, $\text{del}(x_1)$ and $\text{del}(y_1)$ stand for deleting the first elements of $\vec{x}$ and $\vec{y}$, respectively, and sub $(x_1, y_1)$ stands for substituting the first element of $\vec{x}$ with the first element of $\vec{y}$.

A distance function with time warping allows non-uniform scaling along the time axis, or, in sequence terms, *stuttering*. Stuttering occurs when an element from one of the sequences is repeated several times. A typical distance measure is:

$$d_{tw}(\vec{x}, \vec{y}) = d_0(x_1, y_1) + \min \begin{cases} d_{tw}(\vec{x}, y_{2:n}) & (\vec{x} - \text{stutter}), \\ d_{tw}(x_{2:m}, \vec{y}) & (\vec{y} - \text{stutter}), \\ d_{tw}(x_{2:m}, y_{2:n}) & (\text{no stutter}). \end{cases} \tag{8}$$

Both $d_{ed}$ and $d_{tw}$ can be computed in quadratic time ($O(mn)$) using dynamic programming [Cormen *et al.* (1993), Sankoff and Kruskal (1999)]: An $m \times n$ table $D$ is filled iteratively so that $D[i, j] = d(x_{1:i}, y_{1:j})$. The final distance $d(\vec{x}, \vec{y})$, is found in $D[m, n]$.

The *Longest Common Subsequence* (LCS) measure [Cormen *et al.* (1993)], $d_{lcs}(\vec{x}, \vec{y})$, is the length of the longest sequence $\vec{s}$ which is a (possibly non-contiguous) subsequence of both $\vec{x}$ and $\vec{y}$, in other words:

$$d_{lcs}(\vec{x}, \vec{y}) = \max\left\{|\vec{s}|\,\middle|\,\vec{s} \subseteq \vec{x}, \vec{s} \subseteq \vec{y}\right\}. \tag{9}$$

In some applications the measure is normalised by dividing by $\max(|\vec{x}|, |\vec{y}|)$, giving a distance in the range $[0, 1]$. $d_{lcs}(\vec{x}, \vec{y})$ may be calculated using dynamic programming, in a manner quite similar to $d_{ed}$.

## References

1. Agrawal, R., Faloutsos, C., and Swami, A.N. (1993). Efficient Similarity Search in Sequence Databases. *Proc. 4th Int. Conf. on Foundations of Data Organization and Algorithms (FODO)*, pp. 69–84.
2. Agrawal, R., Lin, K., Sawhney, H.S., and Shim, K. (1995). Fast Similarity Search in the Presence of Noise, Scaling, and Translation in Time-Series Database. *Proc. 21st Int. Conf. on Very Large Databases (VLDB)*, pp. 490–501.
3. Baeza-Yates, R. and Gonnet, G.H. (1999). A Fast Algorithm on Average for All-Against-All Sequence Matching. *Proc. 6th String Processing and Information Retrieval Symposium (SPIRE)*, pp. 16–23.
4. Baeza-Yates, R. and Ribeiro-Neto, B. (1999). *Modern Information Retrieval.* ACM Press/Addison–Wesley Longman Limited.
5. Chakrabarti, K. and Mehrotra, S. (1999). The Hybrid Tree: An Index Structure for High Dimensional Feature Spaces. *Proc. 15th Int. Conf. on Data Engineering (ICDE)*, pp. 440–447.
6. Chan, K. and Fu, A.W. (1999). Efficient Time Series Matching by Wavelets. *Proc. 15th Int. Conf. on Data Engineering (ICDE)*, pp. 126–133.
7. Cormen, T.H., Leiserson, C.E., and Rivest, R.L. (1993). *Introduction to Algorithms*, MIT Press.
8. Das, G., Lin, K., Mannila, H., Renganathan, G., and Smyth, P. (1998). Rule Discovery from Time Series. *Proc. 4th Int. Conf. on Knowledge Discovery and Data Mining (KDD)*, pp. 16–22.
9. Faloutsos, C., Jagadish, H.V., Mendelzon, A.O., and Milo, T. (1997). A Signature Technique for Similarity-Based Queries. *Proc. Compression and Complexity of Sequences (SEQUENCES)*.
10. Faloutsos, C. and Lin, K.-I. (1995). FastMap: A Fast Algorithm for Indexing, Data-Mining and Visualization of Traditional and Multimedia Datasets. *Proc. of the 1995 ACM SIGMOD Int. Conf. on Management of Data*, pp. 163–174.
11. Faloutsos, C., Ranganathan, M., and Manolopoulos, Y. (1994). Fast Subsequence Matching in Time-Series Databases. *Proc. of the 1994 ACM SIGMOD Int. Conf. on Management of Data*, pp. 419–429.
12. Gusfield, D. (1997). *Algorithms on Strings, Trees and Sequences*, Cambridge University Press.
13. Guttman, A. (1984). *R*-trees: A Dynamic Index Structure for Spatial Searching. *Proc. 1984 ACM SIGMOD Int. Conf. on Management of Data*, pp. 47–57.
14. Jagadish, H.V., Mendelzon, A.O., and Milo, T. (1995). Similarity-Based Queries. *Proc. 14th Symposium on Principles of Database Systems (PODS)*, pp. 36–45.
15. Keogh, E.J. (1997). A Fast and Robust Method for Pattern Matching in Time Series Databases. *Proc. 9th Int. Conf. on Tools with Artificial Intelligence (ICTAI)*, pp. 578–584.

16. Keogh, E.J. (2002). Exact Indexing of Dynamic Time Warping. *Proc. 28th Int. Conf. on Very Large Data Bases (VLDB)*, pp. 406–417.
17. Keogh, E.J. and Pazzani, M.J. (1998). An Enhanced Representation of Time Series which Allows Fast and Accurate Classification, Clustering and Relevance Feedback. *Proc. 4th Int. Conf. on Knowledge Discovery and Data Mining (KDD)*, pp. 239–243.
18. Keogh, E.J. and Pazzani, M.J. (1999a). An Indexing Scheme for Fast Similarity Search in Large Time Series Databases. *Proc. 11th Int. Conf. on Scientific and Statistical Database Management (SSDBM)*, pp. 56–67.
19. Keogh, E.J. and Pazzani, M.J. (1999b). Scaling up Dynamic Time Warping to Massive Datasets. *Proc. 3rd European Conf. on Principles of Data Mining and Knowledge Discovery (PKDD)*, pp. 1–11.
20. Keogh, E.J. and Pazzani, M.J. (2000). A Simple Dimensionality Reduction Technique for Fast Similarity Search in Large Time Series Databases. *Proc. of the 4th Pacific-Asia Conference on Knowledge Discovery and Data Mining (PAKDD)*, pp. 122–133.
21. Keogh, E.J. and Smyth, P. (1997). A Probabilistic Approach to Fast Pattern Matching in Time Series Databases. *Proc. 3rd Int. Conf. on Knowledge Discovery and Data Mining (KDD)*, pp. 24–30.
22. Keogh, E.J., Chakrabarti, K., Mehrotra, S., and Pazzani, M.J. (2001a). Locally Adaptive Dimensionality Reduction for Indexing Large Time Series Databases. *Proc. 2001 ACM SIGMOD Conf. on Management of Data*, pp. 151–162.
23. Keogh, E.J., Chakrabarti, K., Pazzani, M.J., and Mehrotra, S. (2001b). Dimensionality Reduction for Fast Similarity Search in Large Time Series Databases. *Journal of Knowledge and Information Systems* **3**(3), 263–286.
24. Keogh, E.J., Lonardi, S., and Chiu, B. (2002). Finding Surprising Patterns in a Time Series Database in Linear Time and Space. *Proc. 8th ACM SIGKDD Int. Conf. on Knowledge Discovery and Data Mining (KDD)*, pp. 550–556.
25. Kim, S., Park, S., and Chu, W.W. (2001). An Index-Based Approach for Similarity Search Supporting Time Warping in Large Sequence Databases. *Proc. 17th Int. Conf. on Data Engineering (ICDE)*, pp. 607–614.
26. Lee, S., Chun, S., Kim, D., Lee, J., and Chung, C. (2000). Similarity Search for Multidimensional Data Sequences. *Proc. 16th Int. Conf. on Data Engineering (ICDE)*, pp. 599–609.
27. Mannila, H. and Ronkainen, P. (1997). Similarity of Event Sequences. *Proc. 4th Int. Workshop on Temporal Representation and Reasoning*, TIME, pp. 136–139.
28. Park, S., Chu, W.W., Yoon, J., and Hsu, C. (2000). Efficient Search for Similar Subsequences of Different Lengths in Sequence Databases. *Proc. 16th Int. Conf. on Data Engineering (ICDE)*, pp. 23–32.
29. Perng, C., Wang, H., Zhang, S.R., and Parker, D.S. (2000). Landmarks: A New Model for Similarity-Based Pattern Querying in Time Series Databases. *Proc. 16th Int. Conf. on Data Engineering (ICDE)*, pp. 33–42.

30. Rafiei, D. and Mendelzon, A. (1997). On Similarity-Based Queries for Time Series Data. *SIGMOD Record*, **26**(2), 13–25.
31. Sankoff, D. and Kruskal, J., ed. (1999). *Time Warps, String Edits, and Macromolecules: The Theory and Practice of Sequence Comparison.* CSLI Publications, reissue edition.
32. Sellis, T.K., Roussopoulos, N., and Faloutsos, C. (1987). The $R^+$-Tree: A Dynamic Index for Multi-Dimensional Objects. *Proc. 13th Int. Conf. on Very Large Database (VLDB)*, pp. 507–518.
33. Shatkay, H. (1995). *The fourier transform: A primer.* Technical Report CS-95-37, Brown University.
34. Wang, H. and Perng, C. (2001). The $S^2$-Tree. An Index Structure for Subsequence Matching of Spatial Objects. *Proc. of the 5th Pacific-Asia Conf. on Knowledge Discovery and Data Mining (PAKDD)*, pp. 312–323.
35. Wang, J.T.-L., Wang, X., Lin, K.-I., Shasha, D., Shapiro, B.A., and Zhang, K. (1999). Evaluating a Class of Distance-Mapping Algorithms for Data Mining and Clustering. *Proc. of the 1999 ACM SIGKDD Int. Conf. on Knowledge Discovery and Data Mining*, pp. 307–311.
36. Weber, R., Schek, H., and Blott, S. (1998). A Quantitative Analysis and Performance Study for Similarity-Search Methods in High-Dimensional Spaces. *Proc. 24th Int. Conf. on Very Large Databases (VLDB)*, pp. 194–205.
37. Wu, Y., Agrawal, D., and Abbadi, A.E. (2000). A Comparison of DFT and DWT Based Similarity Search in Time-Series Databases. *Proc. 9th Int. Conf. on Information and Knowledge Management (CIKM)*, pp. 488–495.
38. Yi, B. and Faloutsos, C. (2000). Fast Time Sequence Indexing for Arbitrary $L_p$ Norms. *The VLDB Journal*, pp. 385–594.
39. Yi, B., Jagadish, H.V., and Faloutsos, C. (1998). Efficient Retrieval of Similar Time Sequences Under Time Warping. *Proc. 14th Int. Conf. on Data Engineering (ICDE)*, pp. 201–208.

# CHAPTER 3

# INDEXING OF COMPRESSED TIME SERIES

Eugene Fink
*Computer Science and Engineering, University of South Florida*
*Tampa, Florida 33620, USA*
E-mail: eugene@csee.usf.edu

Kevin B. Pratt
*Harris Corporation, 1025 NASA Blvd., W3/9708*
*Melbourne, Florida 32919, USA*
E-mail: kpratt01@harris.com

We describe a procedure for identifying major minima and maxima of a time series, and present two applications of this procedure. The first application is fast compression of a series, by selecting major extrema and discarding the other points. The compression algorithm runs in linear time and takes constant memory. The second application is indexing of compressed series by their major extrema, and retrieval of series similar to a given pattern. The retrieval procedure searches for the series whose compressed representation is similar to the compressed pattern. It allows the user to control the trade-off between the speed and accuracy of retrieval. We show the effectiveness of the compression and retrieval for stock charts, meteorological data, and electroencephalograms.

*Keywords*: Time series; compression; fast retrieval; similarity measures.

## 1. Introduction

We view a *time series* as a sequence of values measured at equal intervals; for example, the series in Figure 1 includes the values 20, 22, 25, 22, and so on. We describe a compression procedure based on extraction of certain important minima and maxima from a series. For example, we can compress the series in Figure 1 by extracting the circled minima and maxima, and discarding the other points. We also propose a measure of similarity between

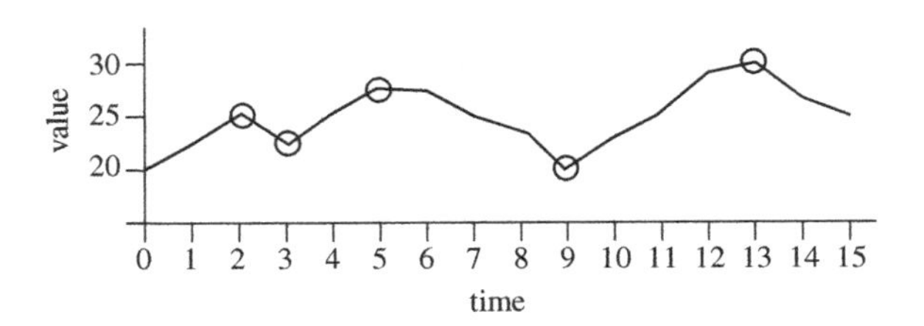


Fig. 1. Example of a time series.

Table 1. Data sets used in the experiments.

| Data set | Description | Number of series | Points per series | Total number of points |
|---|---|---|---|---|
| Stock prices | 98 stocks, 2.3 years | 98 | 610 | 60,000 |
| Air and sea Temperatures | 68 buoys, 18 years, 2 sensors per buoy | 136 | 1,800–6,600 | 450,000 |
| Wind speeds | 12 stations, 18 years | 12 | 6,570 | 79,000 |
| EEG | 64 electrodes, 1 second | 64 | 256 | 16,000 |

series and show that it works well with compressed data. Finally, we present a technique for indexing and retrieval of compressed series; we have tested it on four data sets (Table 1), which are publicly available through the Internet.

*Stock prices:* We have used stocks from the Standard and Poor's 100 listing of large companies for the period from January 1998 to April 2000. We have downloaded daily prices from America Online, discarded newly listed and de-listed stocks, and used ninety-eight stocks in the experiments.

*Air and sea temperatures:* We have experimented with daily temperature readings by sixty-eight buoys in the Pacific Ocean, from 1980 to 1998, downloaded from the Knowledge Discovery archive at the University of California at Irvine (kdd.ics.uci.edu).

*Wind speeds:* We have used daily wind speeds from twelve sites in Ireland, from 1961 to 1978, obtained from an archive at Carnegie Mellon University (lib.stat.cmu.edu/datasets).

*Electroencephalograms:* We have utilized EEG obtained by Henri Begleiter at the Neurodynamics Laboratory of the SUNY Health Center at Brooklyn. These data are from sixty-four electrodes at standard points on the scalp; we have downloaded them from an archive at the University of California at Irvine (kdd.ics.uci.edu).

## 2. Previous Work

We review related work on the comparison and indexing of time series.

**Feature sets.** Researchers have considered various feature sets for compressing time series and measuring similarity between them. They have extensively studied Fourier transforms, which allow fast compression [Singh and McAtackney (1998), Sheikholeslami *et al.* (1998), Stoffer (1999), Yi *et al.* (2000)], however, this technique has several disadvantages. In particular, it smoothes local extrema, which may lead to a loss of important information, and it does not work well for erratic series [Ikeda *et al.* (1999)]. Chan and his colleagues applied Haar wavelet transforms to time series and showed several advantages of this technique over Fourier transforms [Chan and Fu 1999, Chan*et al.* (2003)].

Guralnik and Srivastava (1999) considered the problem of detecting a change in the trend of a data stream, and developed a technique for finding "change points" in a series. Last *et al.* (2001) proposed a general framework for knowledge discovery in time series, which included representation of a series by its key features, such as slope and signal-to-noise ratio. They described a technique for computing these features and identifying the points of change in the feature values.

Researchers have also studied the use of small alphabets for compression of time series, and applied string matching to the pattern search [Agrawal *et al.* (1995), Huang and Yu (1999), Andr-Jnsson and Badal (1997), Lam and Wong (1998), Park *et al.* (1999), Qu *et al.* (1998)]. For example, Guralnik *et al.* (1997) compressed stock prices using a nine-letter alphabet. Singh and McAtackney (1998) represented stock prices, particle dynamics, and stellar light intensity using a three-letter alphabet. Lin and Risch (1998) used a two-letter alphabet to encode major spikes in a series. Das *et al.* (1998) utilized an alphabet of primitive shapes for efficient compression. These techniques give a high compression rate, but their descriptive power is limited, which makes them inapplicable in many domains.

Perng *et al.* (2000) investigated a compression technique based on extraction of "landmark points," which included local minima and maxima. Keogh and Pazzani (1997; 1998) used the endpoints of best-fit line segments to compress a series. Keogh *et al.* (2001) reviewed the compression techniques based on approximation of a time series by a sequence of straight segments. We describe an alternative compression technique, based on selection of important minima and maxima.

**Similarity measures.** Several researchers have defined similarity as the distance between points in a feature space. For example, Caraca-Valente and Lopez-Chavarrias (2000) used Euclidean distance between feature vectors containing angle of knee movement and muscle strength, and Lee *et al.* (2000) applied Euclidean distance to compare feature vectors containing color, texture, and shape of video data. This technique works well when all features have the same units of scale [Goldin and Kanellakis (1995)], but it is often ineffective for combining disparate features.

An alternative definition of similarity is based on bounding rectangles; two series are similar if their bounding rectangles are similar. It allows fast pruning of clearly dissimilar series [Perng *et al.* (2000), Lee *et al.* (2000)], but it is less effective for selecting the most similar series.

The envelope-count technique is based on dividing a series into short segments, called envelopes, and defining a yes/no similarity for each envelope. Two series are similar within an envelope if their point-by-point differences are within a certain threshold. The overall similarity is measured by the number of envelopes where the series are similar [Agrawal *et al.* (1996)]. This measure allows fast computation of similarity, and it can be adapted for noisy and missing data [Das *et al.* (1997), Bollobas *et al.* (1997)].

Finally, we can measure a point-by-point similarity of two series and then aggregate these measures, which often requires interpolation of missing points. For example, Keogh and Pazzani (1998) used linear interpolation with this technique, and Perng *et al.* (2000) applied cubic approximation. Keogh and Pazzani (2000) also described a point-by-point similarity with modified Euclidean distance, which does not require interpolation.

**Indexing and retrieval.** Researchers have studied a variety of techniques for indexing of time series. For example, Deng (1998) applied *kd*-trees to arrange series by their significant features, Chan and Fu (1999) combined wavelet transforms with R-trees, and Bozkaya and her colleagues used vantage-point trees for indexing series by numeric features [Bozkaya *et al.* (1997), Bozkaya and Özsoyoglu (1999)]. Park *et al.* (2001) indexed series by their local extrema and by properties of the segments between consecutive extrema. Li *et al.* (1998) proposed a retrieval technique based on a multi-level abstraction hierarchy of features. Aggarwal and Yu (2000) considered grid structures, but found that the grid performance is often no better than exhaustive search. They also showed that exhaustive search among compressed series is often faster than sophisticated indexing techniques.

## 3. Important Points

We compress a time series by selecting some of its minima and maxima, and dropping the other points (Figure 2). The intuitive idea is to discard minor fluctuations and keep major minima and maxima. We control the compression rate with a parameter $R$, which is always greater than one; an increase of $R$ leads to selection of fewer points. A point $a_m$ of a series $a_1, \ldots, a_n$ is an *important minimum* if there are indices $i$ and $j$, where $i \leq m \leq j$, such that

- $a_m$ is the minimum among $a_i, \ldots, a_j$, and
- $a_i/a_m \geq R$ and $a_j/a_m \geq R$.

Intuitively, $a_m$ is the minimal value of some segment $a_i, \ldots, a_j$, and the end-point values of this segment are much larger than $a_m$ (Figure 3). Similarly, $a_m$ is an *important maximum* if there are indices $i$ and $j$, where $i \leq m \leq j$, such that

- $a_m$ is the minimum among $a_i, \ldots, a_j$, and
- $a_m/a_i \geq R$ and $a_m/a_j \geq R$.

In Figure 4, we give a procedure for selecting important points, which takes linear time and constant memory. It outputs the values and indices of all important points, as well as the first and last point of the series. This procedure can process new points as they arrive, without storing the

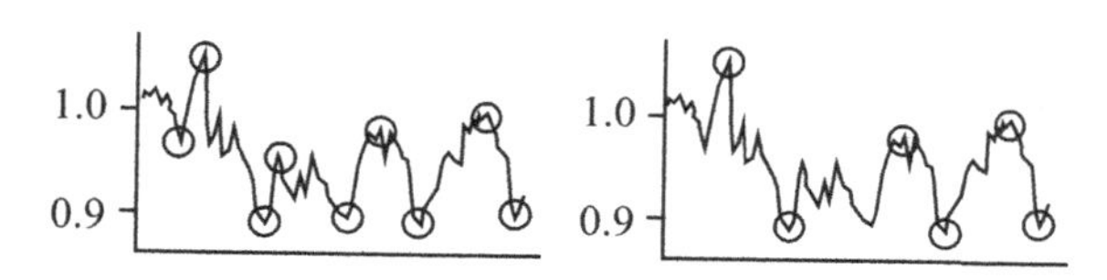


Fig. 2. Important points for 91% compression (left) and 94% compression (right).

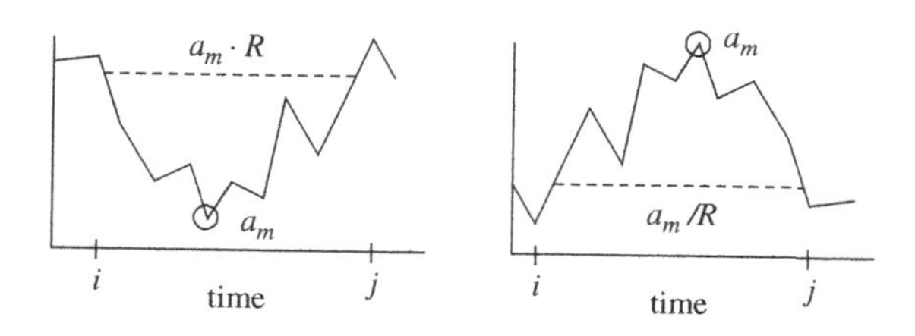


Fig. 3. Important minimum (left) and important maximum (right).

```
IMPORTANT-POINTS—Top-level function for finding important points.
The input is a time series a_1, ..., a_n; the output is the
values and indices of the selected important points.

output (a_1, 1)
i = FIND-FIRST
if i < n and a_i > a_1 then i = FIND-MAXIMUM(i)
while i < n do
  i = FIND-MINIMUM(i)
  i = FIND-MAXIMUM(i)
output (a_n, n)
```

```
FIND-FIRST—Find the first important point.
iMin = 1; iMax = 1; i = 2
while i < n and a_i/a_iMin < R and a_iMax/a_i < R do
  if a_i < a_iMin then iMin = i
  if a_i > a_iMax then iMax = i
  i = i + 1
if iMin < iMax
  then output (a_iMin, iMin)
  else output (a_iMax, iMax)
return i
```

```
FIND-MINIMUM(i)—Find the first important maximum after the ith point.
iMin = i
while i < n and a_i/a_iMin < R do
  if a_i < a_iMin then iMin = i
  i = i + 1
if i < n or a_iMin < a_i then output (a_iMin, iMin)
return i
```

```
FIND-MAXIMUM(i)—Find the first important maximum after the ith point.
iMax = i
while i < n and a_iMax/a_i < R do
  if a_i > a_iMax then iMax = i
  i = i + 1
if i < n or a_iMax > a_i then output (a_iMax, iMax)
return i
```

Fig. 4. Compression procedure. We process a series $a_1, \ldots, a_n$ and use a global variable $n$ to represent its size. The procedure outputs the values and indices of the selected points.

original series; for example, it can compress a live electroencephalogram without waiting until the end of the data collection. We have implemented it in Visual Basic 6 and tested on a 300-MHz PC; for an $n$-point series, the compression time is $14 \cdot n$ microseconds.

We have applied the compression procedure to the data sets in Table 1, and compared it with two simple techniques: equally spaced points and randomly selected points. We have experimented with *different compression rates*, which are defined as the percentage of points removed from a series. For example, "eighty-percent compression" means that we select 20% of points and discard the other 80%.

For each compression technique, we have measured the difference between the original series and the compressed series. We have considered three measures of difference between the original series, $a_1, \ldots, a_n$, and the series interpolated from the compressed version, $b_1, \ldots, b_n$.

Mean difference: $\frac{1}{n} \cdot \sum_{i=1}^{n} |a_i - b_i|$.

Maximum difference: $\max_{i \in [1, \ldots, n]} |a_i - b_i|$.

Root mean square difference: $\sqrt{\frac{1}{n} \cdot \sum_{i=1}^{n} (a_i - b_i)^2}$.

We summarize the results in Table 2, which shows that important points are significantly more accurate than the two simple methods.

## 4. Similarity Measures

We define similarity between time series, which underlies the retrieval procedure. We measure similarity on a zero-to-one scale; zero means no likeness and one means perfect likeness. We review three basic measures of similarity and then propose a new measure. First, we define similarity between two numbers, $a$ and $b$:

$$\text{sim}(a, b) = 1 - \frac{|a - b|}{|a| + |b|}.$$

The *mean similarity* between two series, $a_1, \ldots, a_n$ and $b_1, \ldots, b_n$, is the mean of their point-by-point similarity:

$$\frac{1}{n} \cdot \sum_{i=1}^{n} \text{sim}(a_i, b_i).$$

We also define the *root mean square similarity:*

$$\sqrt{\frac{1}{n} \cdot \sum_{i=1}^{n} \text{sim}(a_i, b_i)^2}.$$

In addition, we consider the correlation coefficient, which is a standard statistical measure of similarity. It ranges from $-1$ to 1, but we can convert

Table 2. Accuracy of three compression techniques. We give the average difference between an original series and its compressed version using the three difference measures; smaller differences correspond to more accurate compression.

| | Mean difference | | | Maximum difference | | | Root mean square diff. | | |
|---|---|---|---|---|---|---|---|---|---|
| | important points | fixed points | random points | important points | fixed points | random points | important points | fixed points | random points |
| *Eighty percent compression* | | | | | | | | | |
| Stock prices | 0.02 | 0.03 | 0.04 | 0.70 | 1.70 | 1.60 | 0.05 | 0.14 | 0.14 |
| Air temp. | 0.01 | 0.03 | 0.03 | 0.33 | 0.77 | 0.72 | 0.03 | 0.10 | 0.10 |
| Sea temp. | 0.01 | 0.03 | 0.03 | 0.35 | 0.81 | 0.75 | 0.03 | 0.10 | 0.10 |
| Wind speeds | 0.02 | 0.03 | 0.03 | 0.04 | 1.09 | 1.01 | 0.04 | 0.05 | 0.05 |
| EEG | 0.03 | 0.06 | 0.07 | 0.68 | 1.08 | 1.00 | 0.10 | 0.18 | 0.17 |
| *Ninety percent compression* | | | | | | | | | |
| Stock prices | 0.03 | 0.06 | 0.07 | 1.10 | 1.70 | 1.70 | 0.08 | 0.21 | 0.21 |
| Air temp. | 0.02 | 0.05 | 0.05 | 0.64 | 0.80 | 0.78 | 0.08 | 0.16 | 0.14 |
| Sea temp. | 0.01 | 0.04 | 0.05 | 0.60 | 0.83 | 0.82 | 0.07 | 0.16 | 0.14 |
| Wind speeds | 0.03 | 0.04 | 0.04 | 0.06 | 1.09 | 1.03 | 0.05 | 0.06 | 0.06 |
| EEG | 0.08 | 0.13 | 0.12 | 0.82 | 1.10 | 1.09 | 0.17 | 0.27 | 0.24 |
| *Ninety-five percent compression* | | | | | | | | | |
| Stock prices | 0.05 | 0.10 | 0.12 | 1.30 | 1.80 | 1.80 | 0.11 | 0.32 | 0.30 |
| Air temp. | 0.03 | 0.09 | 0.08 | 0.74 | 0.83 | 0.83 | 0.12 | 0.23 | 0.21 |
| Sea temp. | 0.03 | 0.08 | 0.08 | 0.78 | 0.85 | 0.85 | 0.12 | 0.23 | 0.21 |
| Wind speeds | 0.05 | 0.04 | 0.04 | 0.08 | 1.09 | 1.10 | 0.07 | 0.08 | 0.08 |
| EEG | 0.13 | 0.17 | 0.16 | 0.90 | 1.10 | 1.10 | 0.24 | 0.31 | 0.28 |

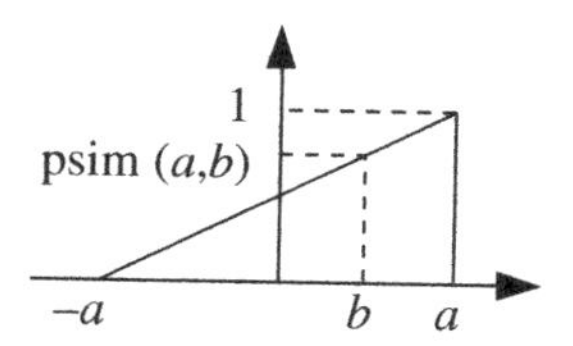


Fig. 5. Peak similarity of numbers $a$ and $b$, where $|a| \geq |b|$.

it to the zero-to-one scale by adding one and dividing by two. For series $a_1, \ldots, a_n$ and $b_1, \ldots, b_n$, with mean values $m_a = (a_1 + \cdots + a_n)/n$ and $m_b = (b_1 + \cdots + b_n)/n$, the correlation coefficient is

$$\frac{\sum_{i=1}^{n}(a_i - m_a) \cdot (b_i - m_b)}{\sqrt{\sum_{i=1}^{n}(a_i - m_a)^2 \cdot \sum_{i=1}^{n}(b_i - m_b)^2}}$$

We next define a new similarity measure, called the *peak similarity*. For numbers $a$ and $b$, their peak similarity is

$$\text{psim}(a, b) = 1 - \frac{|a - b|}{2 \cdot \max(|a|, |b|)}.$$

In Figure 5, we show the meaning of this definition for $|a| \geq |b|$, based on the illustrated triangle. We draw the vertical line through $b$ to the intersection with the triangle's side; the ordinate of the intersection is the similarity of $a$ and $b$. The peak similarity of two series is the mean similarity of their points $(\text{psim}(a_1, b_1) + \cdots + \text{psim}(a_n, b_n))/n$.

We next give an empirical comparison of the four similarity measures. For each series, we have found the five most similar series, and then determined the mean difference between the given series and the other five. In Table 3, we summarize the results and compare them with the perfect exhaustive-search selection and with random selection. The results show that the peak similarity performs better than the other measures, and that the correlation coefficient is the least effective.

We have also used the four similarity measures to identify close matches for each series, and compared the results with ground-truth neighborhoods. For stocks, we have considered small neighborhoods formed by industry subgroups, as well as large neighborhoods formed by industry groups, according to Standard and Poor's classification. For air and sea temperatures, we have used geographic proximity to define two ground-truth neighborhoods. The first neighborhood is the $1 \times 5$ rectangle in the grid of buoys, and the second is the $3 \times 5$ rectangle. For wind speeds, we have also used geographic proximity; the first neighborhood includes all sites within 70 miles,

Table 3. Differences between selected similar series. For each given series, we have selected the five most similar series, and then measured the mean difference between the given series and the other five. Smaller differences correspond to better selection. We also show the running times of selecting similar series.

| Similarity measure | Comp. rate | Stock prices | | | Sea temperatures | | | Air temperatures | | | Wind speeds | | | EEG | | |
|---|---|---|---|---|---|---|---|---|---|---|---|---|---|---|---|---|
| | | Mean diff. | Max diff. | Time (sec) | Mean diff. | Max diff. | Time (sec) | Mean diff. | Max diff. | Time (sec) | Mean diff. | Max diff. | Time (sec) | Mean diff. | Max diff. | Time (sec) |
| Perfect selection | | 0.094 | 0.437 | | 0.016 | 0.072 | | 0.024 | 0.121 | | 0.021 | 0.136 | | 0.038 | 0.170 | |
| Random selection | | 0.287 | 1.453 | | 0.078 | 0.215 | | 0.070 | 0.235 | | 0.029 | 0.185 | | 0.072 | 0.370 | |
| Peak | 90% | 0.103 | 0.429 | 0.024 | 0.018 | 0.068 | 0.021 | 0.029 | 0.103 | 0.022 | 0.023 | 0.138 | 0.016 | 0.052 | 0.241 | 0.015 |
| Similarity | 95% | 0.110 | 0.534 | 0.022 | 0.019 | 0.073 | 0.019 | 0.030 | 0.136 | 0.020 | 0.023 | 0.148 | 0.016 | 0.063 | 0.306 | 0.015 |
| Mean | 90% | 0.110 | 0.525 | 0.026 | 0.026 | 0.092 | 0.022 | 0.031 | 0.134 | 0.022 | 0.023 | 0.137 | 0.017 | 0.055 | 0.279 | 0.016 |
| Similarity | 95% | 0.126 | 0.570 | 0.024 | 0.033 | 0.112 | 0.021 | 0.037 | 0.152 | 0.022 | 0.025 | 0.152 | 0.017 | 0.066 | 0.323 | 0.014 |
| Root mean | 90% | 0.103 | 0.497 | 0.026 | 0.024 | 0.090 | 0.022 | 0.030 | 0.133 | 0.022 | 0.023 | 0.134 | 0.017 | 0.051 | 0.261 | 0.016 |
| Square sim. | 95% | 0.115 | 0.588 | 0.024 | 0.031 | 0.106 | 0.021 | 0.035 | 0.147 | 0.022 | 0.023 | 0.153 | 0.017 | 0.064 | 0.317 | 0.014 |
| Correlation | 90% | 0.206 | 1.019 | 0.048 | 0.054 | 0.162 | 0.044 | 0.051 | 0.214 | 0.046 | 0.024 | 0.138 | 0.042 | 0.056 | 0.281 | 0.030 |
| Coefficient | 95% | 0.210 | 1.101 | 0.045 | 0.063 | 0.179 | 0.042 | 0.051 | 0.224 | 0.043 | 0.024 | 0.154 | 0.033 | 0.068 | 0.349 | 0.028 |

and the second includes the sites within 140 miles. For electroencephalograms, the first neighborhood is the $3 \times 3$ rectangle in the grid of electrodes, and the second is the $5 \times 5$ rectangle.

For each series, we have found the five closest matches, and then determined the average number of matches that belong to the same neighborhood. In Table 4, we give the results and compare them with the perfect selection and random selection; larger numbers correspond to better selections.

The results show that the peak similarity is usually more effective than the other three similarities. If we use the 90% compression, the peak similarity gives better selection than the other similarity measures for the stock prices, air and sea temperatures, and EEG; however, it gives worse results for the wind speeds. If we use the 95% compression, the peak similarity outperforms the other measures on the stocks, temperatures, EEG, and large-neighborhood selection for the wind speeds; however, it loses to the mean similarity and correlation coefficient on the small-neighborhood selection for the wind speeds.

We have also checked how well the peak similarity of original series correlates with the peak similarity of their compressed versions (Figure 6). We have observed a good linear correlation, which gracefully degrades with an increase of compression rate.

## 5. Pattern Retrieval

We give an algorithm that inputs a pattern series and retrieves similar series from a database. It includes three steps: identifying a "prominent feature" of the pattern, finding similar features in the database, and comparing the pattern with each series that has a similar feature.

We begin by defining a *leg* of a series, which is the segment between two consecutive important points. For each leg, we store the values listed in Figure 7, denoted *vl, vr, il, ir, ratio*, and *length;* we give an example of these values in Figure 8. The *prominent leg* of a pattern is the leg with the greatest endpoint ratio.

The retrieval procedure inputs a pattern and searches for similar segments in a database (Figure 9). First, it finds the pattern leg with the greatest endpoint ratio, denoted $ratio_p$, and determines the length of this leg, $length_p$. Next, it identifies all legs in the database that have a similar endpoint ratio and length. A leg is considered similar to the pattern leg if its ratio is between $ratio_p/C$ and $ratio_p \cdot C$, and its length is between

Table 4. Finding members of the same neighborhood. For each series, we have found the five closest matches, and then determined the average number of the series among them that belong to the same ground-truth neighborhood.

| Similarity Measure | Comp. rate | Stocks | | Sea temp. | | Air temp. | | Wind speeds | | EEG | |
|---|---|---|---|---|---|---|---|---|---|---|---|
| | | 1 | 2 | 1 | 2 | 1 | 2 | 1 | 2 | 1 | 2 |
| Perfect selection | | 5.00 | 5.00 | 5.00 | 5.00 | 5.00 | 5.00 | 5.00 | 5.00 | 5.00 | 5.00 |
| Random selection | | 0.07 | 0.29 | 0.11 | 0.40 | 0.10 | 0.40 | 0.74 | 2.27 | 0.35 | 1.03 |
| Peak | 90% | 0.22 | 0.62 | 0.54 | 1.09 | 0.49 | 0.89 | 1.16 | 2.83 | 1.81 | 2.81 |
| Similarity | 95% | 0.21 | 0.55 | 0.65 | 1.18 | 0.48 | 0.82 | 1.50 | 2.83 | 1.59 | 2.25 |
| Mean | 90% | 0.18 | 0.55 | 0.28 | 0.85 | 0.34 | 0.77 | 1.33 | 2.92 | 1.05 | 1.98 |
| Similarity | 95% | 0.12 | 0.47 | 0.17 | 0.75 | 0.25 | 0.65 | 1.58 | 2.66 | 0.36 | 0.90 |
| Root mean | 90% | 0.14 | 0.53 | 0.32 | 0.88 | 0.34 | 0.83 | 1.50 | 2.92 | 1.20 | 2.19 |
| Square sim. | 95% | 0.17 | 0.35 | 0.20 | 0.77 | 0.26 | 0.71 | 1.33 | 2.75 | 0.36 | 0.90 |
| Correlation | 90% | 0.15 | 0.39 | 0.25 | 0.82 | 0.49 | 0.74 | 1.33 | 2.92 | 1.16 | 2.24 |
| Coefficient | 95% | 0.19 | 0.50 | 0.29 | 0.72 | 0.34 | 0.60 | 1.51 | 2.75 | 0.68 | 1.65 |

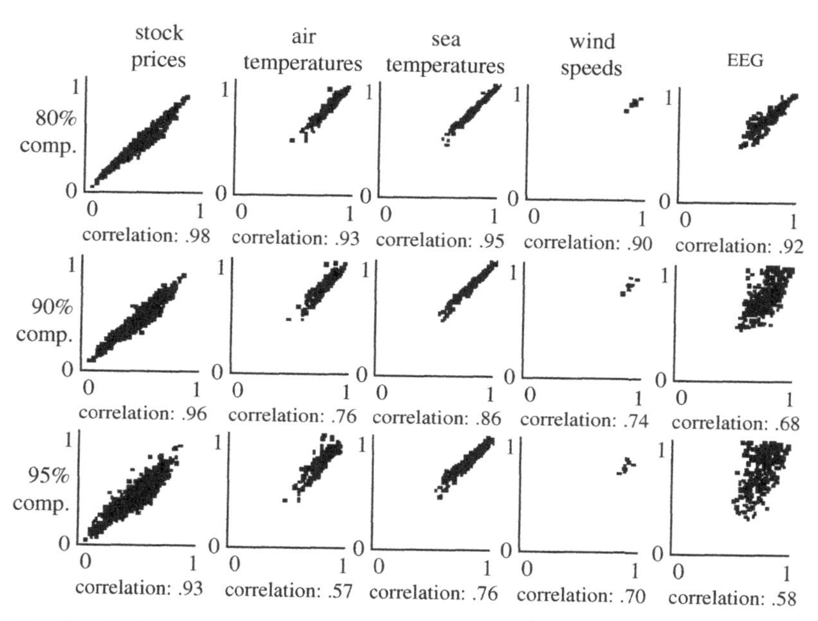


Fig. 6. Correlation between the peak similarity of original series and the peak similarity of their compressed versions. We show the correlation for three compression rates: 80%, 90%, and 95%.

| | |
|---|---|
| *vl* | value of the left important point of the leg |
| *vr* | value of the right important point of the leg |
| *il* | index of the left important point |
| *ir* | index of the right important point |
| *ratio* | ratio of the endpoints, defined as $vr/vl$ |
| *length* | length of the leg, defined as $ir - il$ |

Fig. 7. Basic data for a leg.

$length_p/D$ and $length_p \cdot D$, where $C$ and $D$ are parameters for controlling the search.

We index all legs in the database by their ratio and length using a *range tree,* which is a standard structure for indexing points by two co-ordinates [Edelsbrunner (1981), Samet (1990)]. If the total number of legs is $l$, and the number of retrieved legs with an appropriate ratio and length is $k$, then the retrieval time is $O(k + \lg l)$.

Finally, the procedure identifies the segments that contain the selected legs (Figure 10) and computes their similarity to the pattern. If the similarity is above a given threshold $T$, the procedure outputs the segment as a match. In Figure 11, we give an example of a stock-price pattern and similar segments retrieved from the stock database.

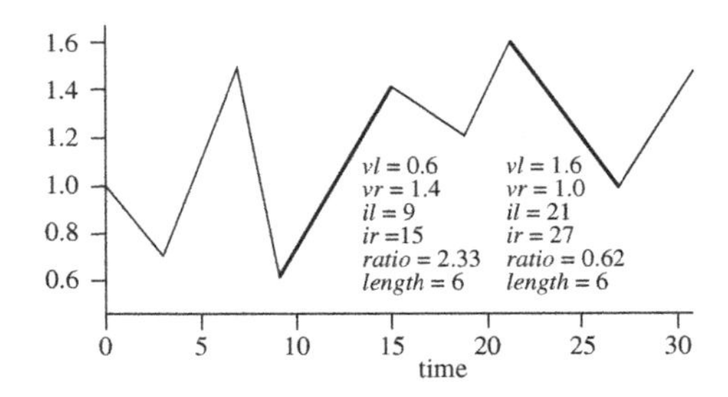


Fig. 8. Example legs. We show the basic data for the two legs marked by thick lines.

---

PATTERN-RETRIEVAL

**The procedure inputs a pattern series and searches a time-series database; the output is a list of segments from the database that match the pattern.**

Identify the pattern leg $p$ with the greatest endpoint ratio, denoted $ratio_p$. Determine the length of this pattern leg, denoted $length_p$.
Find all legs in the database that satisfy the following conditions:

- their endpoint ratios are between $ratio_p/C$ and $ratio_p \cdot C$, and
- their lengths are between $length_p/D$ and $length_p \cdot D$.

For each leg in the set of selected legs:
Identify the segment corresponding to the pattern (Figure 10).
Compute the similarity between the segment and the pattern.
If the similarity is above the threshold $T$, then output the segment.

---

Fig. 9. Search for segments similar to a given pattern. We use three parameters to control the search: maximal ratio deviation $C$, maximal length deviation $D$, and similarity threshold $T$.

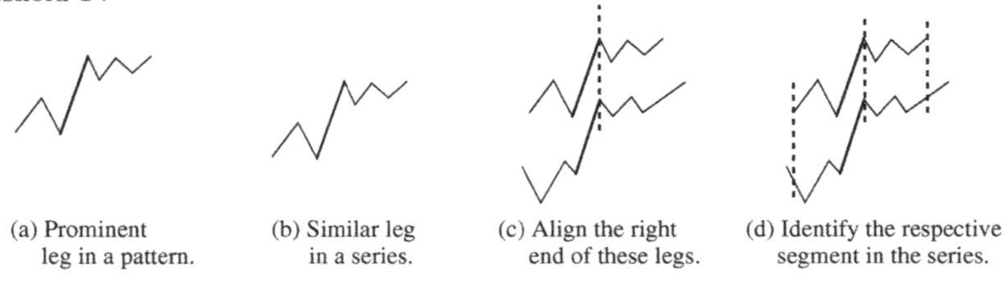


Fig. 10. Identifying a segment that may match the pattern.

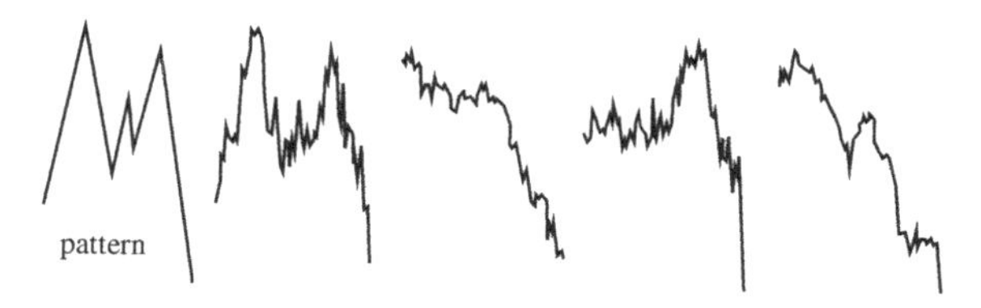


Fig. 11. Example of retrieved stock charts.

Fig. 12. Example of extended legs. The pattern (a) matches the series (b), but the pattern's prominent leg has no equivalent in the series. If we identify the extended legs (c), the prominent leg matches one of them.

The described procedure can miss a matching segment that does not have a leg corresponding to the pattern's prominent leg. We illustrate this problem in Figure 12, where the prominent leg of the pattern has no equivalent in the matching series. To avoid this problem, we introduce the notion of an *extended leg,* which is a segment that would be a leg under a higher compression rate (Figure 12c). Formally, points $a_i$ and $a_j$ of a series $a_1, \ldots, a_n$ form an extended upward leg if

- $a_i$ is a local minimum, and $a_j$ is a local maximum, and
- for every $m \in [i, j]$, we have $a_i < a_m < a_j$.

The definition of an extended downward leg is similar.

We identify all extended legs, and index them in the same way as normal legs. The advantage of this approach is more accurate retrieval, and the disadvantage is a larger indexing structure. In Figure 13, we give an algorithm for identifying upward extended legs; the procedure for finding downward extended legs is similar. We assume that normal upward legs in the input series are numbered from 1 to $l$. First, the procedure processes important maxima; for each maximum $ir_k$, it identifies the *next larger* maximum and stores its index in $next[k]$. Second, it uses this information to identify extended legs. The running time of the first part is linear in the

```
EXTENDED-LEGS
The input is the list of legs in a series; the output is a list of all extended legs.
initialize an empty stack S of leg indices
PUSH(S, 1)
for k = 2 to l do
  while S is not empty and ir_TOP(S) < ir_k do
    next[TOP(S)] = k; POP(S)
  PUSH(k)
while S is not empty do
  next[TOP(S)] = NIL; POP(S)
initialize an empty list of extended legs
for k = 1 to l - 1 do
  m = next[k]
  while m is not NIL do
    add (il_k, ir_m) to the list of extended legs
    m = next[m]
```

Fig. 13. Identifying extended legs. We assume that normal legs are numbered 1 to $l$.

number of normal legs, and the time of the second part is linear in the number of extended legs.

To evaluate the retrieval accuracy, we have compared the retrieval results with the matches identified by a slow exhaustive search. We have ranked the matches found by the retrieval algorithm from most to least similar. In Figures 14 and 15, we plot the ranks of matches found by the fast algorithm versus the ranks of exhaustive-search matches. For instance, if the fast algorithm has found only three among seven closest matches, the graph includes the point (3, 7). Note that the fast algorithm never returns "false positives" since it verifies each candidate match.

The retrieval time grows linearly with the pattern length and with the number of candidate segments identified at the initial step of the retrieval algorithm. If we increase $C$ and $D$, the procedure finds more candidates and misses fewer matchers, but the retrieval takes more time. In Table 5, we give the mean number of candidate segments in the retrieval experiments, for three different values of $C$ and $D$; note that this number does not depend on the pattern length.

We have measured the speed of a Visual-Basic implementation on a 300-MHz PC. If the pattern has $m$ legs, and the procedure identifies $k$ candidate matches, then the retrieval time is $70 \cdot m \cdot k$ microseconds. For the stock

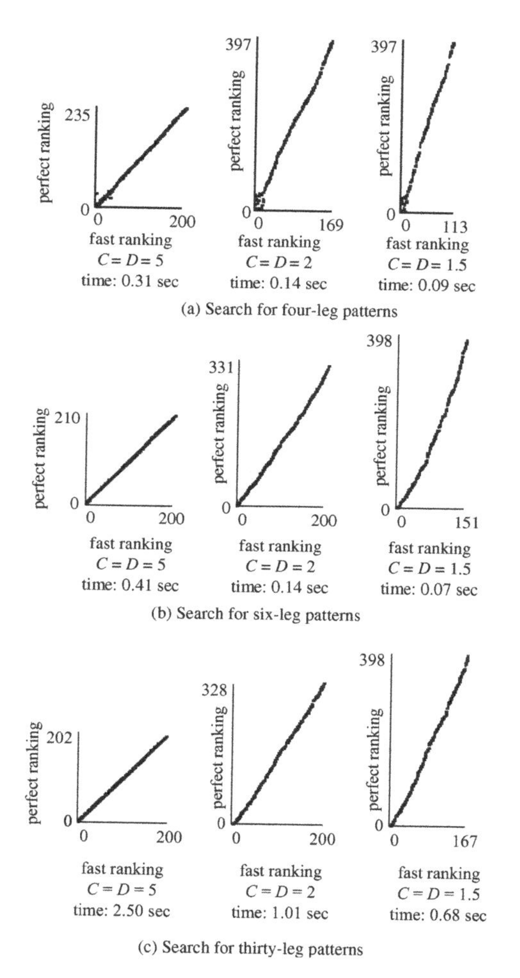


Fig. 14. Retrieval of stock charts. The horizontal axes show the ranks of matches retrieved by the fast algorithm. The vertical axes are the ranks assigned to the same matches by the exhaustive search. If the fast algorithm has found all matches, the graph is a forty-five degree line; otherwise, it is steeper.

database with 60,000 points, the retrieval takes from 0.1 to 2.5 seconds. For the database of air and sea temperatures, which includes 450,000 points, the time is between 1 and 10 seconds.

## 6. Concluding Remarks

The main results include a procedure for compressing time series, indexing of compressed series by their prominent features, and retrieval of series

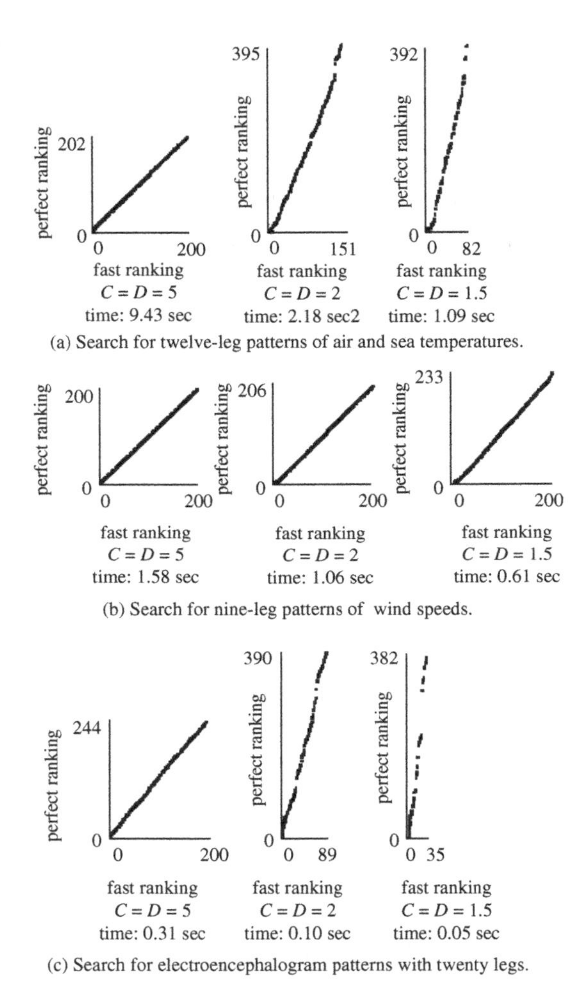


Fig. 15. Retrieval of weather and electroencephalogram patterns. The horizontal axes show the similarity ranks assigned by the fast algorithm, and the vertical axes are the exhaustive-search ranks.

whose compressed representation is similar to the compressed pattern. The experiments have shown the effectiveness of this technique for indexing of stock prices, weather data, and electroencephalograms. We plan to apply it to other time-series domains and study the factors that affect its effectiveness.

We are working on an extended version of the compression procedure, which will assign different importance levels to the extrema of time series,

Table 5. Average number of the candidate segments that match a pattern's prominent leg. The retrieval algorithm identifies these candidates and then compares them with the pattern. The number of candidates depends on the search parameters $C$ and $D$.

| Search parameters | Stock prices | Air and sea temperatures | Wind speeds | EEG |
|---|---|---|---|---|
| $C = D = 1.5$ | 270 | 1,300 | 970 | 40 |
| $C = D = 2$ | 440 | 2,590 | 1,680 | 70 |
| $C = D = 5$ | 1,090 | 11,230 | 2,510 | 220 |

and allow construction of a hierarchical indexing structure [Gandhi (2003)]. We also aim to extend the developed technique for finding patterns that are stretched over time, and apply it to identifying periodic patterns, such as weather cycles.

## Acknowledgements

We are grateful to Mark Last, Eamonn Keogh, Dmitry Goldgof, and Rafael Perez for their valuable comments and suggestions. We also thank Savvas Nikiforou for his comments and help with software installations.

## References

1. Aggarwal, C.C. and Yu, Ph.S. (2000). The IGrid Index: Reversing the Dimensionality Curse for Similarity Indexing in High-Dimensional Space. *Proceedings of the Sixth ACM International Conference on Knowledge Discovery and Data Mining*, pp. 119–129.
2. Agrawal, R., Psaila, G., Wimmers, E.L., and Zait, M. (1995). Querying Shapes of Histories. *Proceedings of the Twenty-First International Conference on Very Large Data Bases*, pp. 502–514.
3. Agrawal, R., Mehta, M., Shafer, J.C., Srikant, R., Arning, A., and Bollinger, T. (1996). The Quest Data Mining System. *Proceedings of the Second ACM International Conference on Knowledge Discovery and Data Mining*, pp. 244–249.
4. André-Jönsson, H. and Badal, D.Z. (1997). Using Signature Files for Querying Time-Series Data. *Proceedings of the First European Symposium on Principles of Data Mining and Knowledge Discovery*, pp. 211–220.
5. Bollobas, B., Das, G., Gunopulos, D., and Mannila, H. (1977). Time-Series Similarity Problems and Well-Separated Geometric Sets. *Proceedings of the Thirteenth Annual Symposium on Computational Geometry*, pp. 454–456.
6. Bozkaya, T. and Özsoyoglu, Z.M. (1999). Indexing Large Metric Spaces for Similarity Search Queries. *ACM Transactions on Database Systems* **24**(3), 361–404.
7. Bozkaya, T., Yazdani, N., and Özsoyoglu, Z.M. (1997). Matching and Indexing Sequences of Different Lengths. *Proceedings of the Sixth International Conference on Information and Knowledge Management*, pp. 128–135.

8. Brockwell, P.J. and Davis, R.A. (1996). *Introduction to Time Series and Forecasting.* Springer-Verlag, Berlin, Germany.
9. Burrus, C.S., Gopinath, R.A., and Guo, H. (1997). *Introduction to Wavelets and Wavelet Transforms: A Primer.* Prentice Hall, Englewood Cliffs, NJ.
10. Caraca-Valente, J.P. and Lopez-Chavarrias, I. (2000). Discovering Similar Patterns in Time Series. *Proceedings of the Sixth ACM International Conference on Knowledge Discovery and Data Mining*, pp. 497–505.
11. Chan, K.-P. and Fu, A.W.-C. (1999). Efficient Time Series Matching by Wavelets. *Proceedings of the Fifteenth International Conference on Data Engineering*, pp. 126–133.
12. Chan, K.-P., Fu, A.W.-C., and Yu, C. (2003). Haar Wavelets for Efficient Similarity Search of Time-Series: With and Without Time Warping. *IEEE Transactions on Knowledge and Data Engineering*, Forthcoming.
13. Chi, E.H.-H., Barry, Ph., Shoop, E., Carlis, J.V., Retzel, E., and Riedl, J. (1995). Visualization of Biological Sequence Similarity Search Results. *Proceedings of the IEEE Conference on Visualization*, pp. 44–51.
14. Cortes, C., Fisher, K., Pregibon, D., and Rogers, A. (2000). Hancock: A Language for Extracting Signatures from Data Streams. *Proceedings of the Sixth ACM International Conference on Knowledge Discovery and Data Mining*, pp. 9–17.
15. Das, G., Gunopulos, D., and Mannila, H. (1997). Finding Similar Time Series. *Proceedings of the First European Conference on Principles of Data Mining and Knowledge Discovery*, pp. 88–100.
16. Das, G., Lin, K-I., Mannila, H., Renganathan, G., and Smyth, P. (1998). Rule Discovery from Time Series. *Proceedings of the Fourth ACM International Conference on Knowledge Discovery and Data Mining*, pp. 16–22.
17. Deng, K. (1998). OMEGA: *On-Line Memory-Based General Purpose System Classifier.* PhD thesis, Robotics Institute, Carnegie Mellon University. Technical Report CMU-RI-TR-98-33.
18. Domingos, P. and Hulten, G. (2000). Mining High-Speed Data Streams. *Proceedings of the Sixth ACM International Conference on Knowledge Discovery and Data Mining*, pp. 71–80.
19. Edelsbrunner, H. (1981). A Note on Dynamic Range Searching. *Bulletin of the European Association for Theoretical Computer Science*, **15**, 34–40.
20. Fountain, T., Dietterich, T.G., and Sudyka, B. (2000). Mining IC Test Data to Optimize VLSI Testing. *Proceedings of the Sixth ACM International Conference on Knowledge Discovery and Data Mining*, pp. 18–25.
21. Franses, P.H. (1998). *Time Series Models for Business and Economic Forecasting.* Cambridge University Press, Cambridge, United Kingdom.
22. Galka, A. (2000). *Topics in Nonlinear Time Series Analysis with Implications for EEG Analysis*, World Scientific, Singapore.
23. Gandhi, H.S. (2003). Important Extrema of Time Series: Theory and Applications. Master's thesis, Computer Science and Engineering, University of South Florida. Forthcoming.

24. Gavrilov, M., Anguelov, D., Indyk, P., and Motwani, R. (2000). Mining the Stock Market: Which Measure is Best. *Proceedings of the Sixth ACM International Conference on Knowledge Discovery and Data Mining*, pp. 487–496.
25. Geva., A.B. (1999). Hierarchical-Fuzzy Clustering of Temporal Patterns and its Application for Time-Series Prediction. *Pattern Recognition Letters*, **20**(14), 1519–1532.
26. Goldin, D.Q., and Kanellakis, P.C. (1995). On Similarity Queries for Time-Series Data: Constraint Specification and Implementation. *Proceedings of the First International Conference on Principles and Practice of Constraint Programming*, pp. 137–153.
27. Graps, A.L. (1995). An Introduction to Wavelets. *IEEE Computational Science and Engineering*, **2**(2), 50–61.
28. Grenander, U. (1996). *Elements of Pattern Theory*, Johns Hopkins University Press, Baltimore, MD.
29. Guralnik, V., and Srivastava, J. (1999). Event Detection from Time Series Data. *Proceedings of the Fifth ACM SIGKDD International Conference on Knowledge Discovery and Data Mining*, pp. 33–42.
30. Guralnik, V., Wijesekera, D., and Srivastava, J. (1998). Pattern-Directed Mining of Sequence Data. *Proceedings of the Fourth ACM International Conference on Knowledge Discovery and Data Mining*, pp. 51–57.
31. Gusfield, D. (1997). *Algorithms on Strings, Trees, and Sequences: Computer Science and Computational Biology*, Cambridge University Press, Cambridge, United Kingdom.
32. Han, J., Gong, W., and Yin, Y. (1998). Mining Segment-Wise Periodic Patterns in Time-Related Databases. *Proceedings of the Fourth ACM International Conference on Knowledge Discovery and Data Mining*, pp. 214–218.
33. Haslett, J., and Raftery, A.E. (1989). Space-Time Modeling with Long-Memory Dependence: Assessing Ireland's Wind Power Resource. *Applied Statistics*, **38**, 1–50.
34. Huang, Y.-W. and Yu, P.S. (1999). Adaptive Query Processing for Time-Series Data. *Proceedings of the Fifth International Conference on Knowledge Discovery and Data Mining*, pp. 282–286.
35. Ikeda, K., Vaughn, B.V., and Quint, S.R. (1999). Wavelet Decomposition of Heart Period Data. *Proceedings of the IEEE First Joint BMES/EMBS Conference*, pp. 3–11.
36. Keogh, E.J. and Pazzani, M.J. (1998). An Enhanced Representation of Time Series which Allows Fast and Accurate Classification, Clustering and Relevance Feedback. *Proceedings of the Fourth ACM International Conference on Knowledge Discovery and Data Mining*, pp. 239–243.
37. Keogh, E.J., and Pazzani, M.J. (2000). Scaling up Dynamic Time Warping for Data Mining Applications. *Proceedings of the Sixth ACM International Conference on Knowledge Discovery and Data Mining*, pp. 285–289.

38. Keogh, E.J., Chu, S., Hart, D., and Pazzani, M.J. (2001). An Online Algorithm for Segmenting Time Series. *Proceedings of the IEEE International Conference on Data Mining*, pp. 289–296.
39. Keogh, E.J. (1997). Fast Similarity Search in the Presence of Longitudinal Scaling in Time Series Databases. *Proceedings of the Ninth IEEE International Conference on Tools with Artificial Intelligence*, pp. 578–584.
40. Kim, S.-W., Park, S., and Chu, W.W. (2001). An Index-Based Approach for Similarity Search Supporting Time Warping in Large Sequence Databases. *Proceedings of the Seventeenth International Conference on Data Engineering*, pp. 607–614.
41. Lam, S.K. and Wong, M.H. (1998). A Fast Projection Algorithm for Sequence Data Searching. *Data and Knowledge Engineering*, **28**(3), 321–339.
42. Last, M., Klein, Y., and Kandel, A. (2001). Knowledge Discovery in Time Series Databases. *IEEE Transactions on Systems, Man, and Cybernetics*, Part B, **31**(1), 160–169.
43. Lee, S.-L., Chun, S.-J., Kim, D.-H., Lee, J.-H., and Chung, C.-W. (2000). Similarity Search for Multidimensional Data Sequences. *Proceedings of the Sixteenth International Conference on Data Engineering*, pp. 599–608.
44. Li, C.-S., Yu, P.S., and Castelli, V. (1998). MALM: A Framework for Mining Sequence Database at Multiple Abstraction Levels. *Proceedings of the Seventh International Conference on Information and Knowledge Management*, pp. 267–272.
45. Lin, L., and Risch, T. (1998). Querying Continuous Time Sequences. *Proceedings of the Twenty-Fourth International Conference on Very Large Data Bases*, pp. 170–181.
46. Lu, H., Han, J., and Feng, L. (1998). Stock Movement Prediction and $N$-Dimensional Inter-Transaction Association Rules. *Proceedings of the ACM SIGMOD Workshop on Research Issues in Data Mining and Knowledge Discovery*, pp. 12:1–7.
47. Maurer, K., and Dierks, T. (1991). *Atlas of Brain Mapping: Topographic Mapping of EEG and Evoked Potentials*, Springer-Verlag, Berlin, Germany.
48. Niedermeyer, E., and Lopes DaSilva, F. (1999). *Electroencephalography: Basic Principles, Clinical Applications, and Related Fields.* Lippincott, Williams and Wilkins, Baltimore, MD, fourth edition.
49. Park, S., Lee, D., and Chu, W.W. (1999). Fast Retrieval of Similar Subsequences in Long Sequence Databases. *Proceedings of the Third IEEE Knowledge and Data Engineering Exchange Workshop.*
50. Park, S., Chu, W.W., Yoon, J., and Hsu, C. (2000). Efficient Searches for Similar Subsequences of Different Lengths in Sequence Databases. *Proceedings of the Sixteenth International Conference on Data Engineering*, pp. 23–32.
51. Park, S., Kim, S.-W., and Chu, W.W. (2001). Segment-Based Approach for Subsequence Searches in Sequence Databases. *Proceedings of the Sixteenth ACM Symposium on Applied Computing*, pp. 248–252.

52. Perng, C.-S., Wang, H., Zhang, S.R., and Scott Parker, D. (2000). Landmarks: A New Model for Similarity-Based Pattern Querying in Time Series Databases. *Proceedings of the Sixteenth International Conference on Data Engineering*, pp. 33–42.
53. Policker, S., and Geva, A.B. (2000). Non-Stationary Time Series Analysis by Temporal Clustering. *IEEE Transactions on Systems, Man, and Cybernetics*, Part B, **30**(2), 339–343.
54. Pratt, K.B. and Fink, E. (2002). Search for Patterns in Compressed Time Series. *International Journal of Image and Graphics*, **2**(1), 89–106.
55. Pratt, K.B. (2001). Locating patterns in discrete time series. Master's thesis, Computer Science and Engineering, University of South Florida.
56. Qu, Y., Wang, C., and Wang, X.S. (1998). Supporting Fast Search in Time Series for Movement Patterns in Multiple Scales. *Proceedings of the Seventh International Conference on Information and Knowledge Management*, pp. 251–258.
57. Sahoo, P.K., Soltani, S., Wong, A.K.C., and Chen, Y.C. (1988). A Survey of Thresholding Techniques. *Computer Vision, Graphics and Image Processing*, **41**, 233–260.
58. Samet, H. (1990). *The Design and Analysis of Spatial Data Structures*. Addison-Wesley, Reading, MA.
59. Sheikholeslami, G., Chatterjee, S., and Zhang, A. (1998). WaveCluster: A Multi-Resolution Clustering Approach for Very Large Spatial Databases. *Proceedings of the Twenty-Fourth International Conference on Very Large Data Bases*, pp. 428–439.
60. Singh, S., and McAtackney, P. (1998). Dynamic Time-Series Forecasting Using Local Approximation. *Proceedings of the Tenth IEEE International Conference on Tools with Artificial Intelligence*, pp. 392–399.
61. Stoffer, D.S. (1999). Detecting Common Signals in Multiple Time Series Using the Spectral Envelope. *Journal of the American Statistical Association*, **94**, 1341–1356.
62. Yaffee, R.A. and McGee, M. (2000). *Introduction to Time Series Analysis and Forecasting*, Academic Press, San Diego, CA.
63. Yi, B.-K, Sidiropoulos, N.D., Johnson, T., Jagadish, H.V., Faloutsos, C., and Biliris, A. (2000). Online Data Mining for Co-Evolving Time Series. *Proceedings of the Sixteenth International Conference on Data Engineering*, pp. 13–22.

# CHAPTER 4

# INDEXING TIME-SERIES UNDER CONDITIONS OF NOISE

Michail Vlachos and Dimitrios Gunopulos
*Department of Computer Science and Engineering*
*Bourns College of Engineering*
*University of California, Riverside Riverside, CA 92521, USA*
E-mail: {mvlachos, dg}@cs.ucr.edu

Gautam Das
*Microsoft Research, One Microsoft Way, Redmond, WA 98052, USA*
E-mail: gautamd@microsoft.com

We present techniques for the analysis and retrieval of time-series under conditions of noise. This is an important topic because the data obtained using various sensors (examples include GPS data or video tracking data) are typically noisy. The performance of previously used measures is generally degraded under noisy conditions. Here we formalize non-metric similarity functions based on the Longest Common Subsequence that are very robust to noise. Furthermore they provide an intuitive notion of similarity between time-series by giving more weight to the similar portions of the sequences. Stretching of sequences in time is allowed, as well as global translating of the sequences in space. Efficient approximate algorithms that compute these similarity measures are also provided. We compare these new methods to the widely used Euclidean and Time Warping distance functions (for real and synthetic data) and show the superiority of our approach, especially under the strong presence of noise. We prove a weaker version of the triangle inequality and employ it in an indexing structure to Answer nearest neighbor queries. Finally, we present experimental results that validate the accuracy and efficiency of our approach.

*Keywords*: Longest Common Subsequence; time-series; spatio-temporal; outliers; time warping.

## 1. Introduction

We consider the problem of discovering similar time-series, especially under the presence of noise. Time-series data come up in a variety of domains, including stock market analysis, environmental data, telecommunication data, medical and financial data. Web data that count the number of clicks on given sites, or model the usage of different pages are also modeled as time series.

In the last few years, the advances in mobile computing, sensor and GPS technology have made it possible to collect large amounts of spatiotemporal data and there is increasing interest to perform data analysis tasks over this data [4]. For example, in mobile computing, users equipped with mobile devices move in space and register their location at different time instants to spatiotemporal databases via wireless links. In environmental information systems, tracking animals and weather conditions is very common and large datasets can be created by storing locations of observed objects over time. Other examples of the data that we are interested in include features extracted from video clips, sign language recognition, and so on.

Data analysis in such data includes determining and finding objects that moved in a similar way or followed a certain motion pattern. Therefore, the objective is to cluster different objects into similar groups, or to classify an object based on a set of known examples. The problem is hard, because the similarity model should allow for imprecise matches.

In general the time-series will be obtained during a tracking procedure, with the aid of devices such as GPS transmitters, data gloves etc. Here also

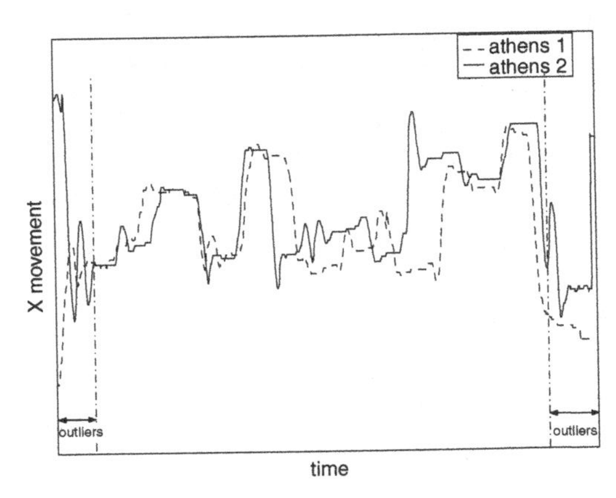


Fig. 1. Examples of video-tracked data representing 2 instances of the word '*athens*'. Start & ending contain many outliers.

lies the main obstacle of such data; they may contain a significant amount of *outliers* (Figure 1) or in other words incorrect data measurements.

We argue that the use of non-metric distance functions is preferable, since such functions can effectively 'ignore' the noisy sections of the sequences.

The rest of the chapter is organized as follows. In section 2 we will review the most prevalent similarity measures used in time-series databases, we will demonstrate the need for non-metric distance functions and we will also consider related work. In section 3 we formalize the new similarity functions by extending the *Longest Common Subsequence (LCSS)* model. Section 4 demonstrates efficient algorithms to compute these functions and section 5 elaborates on the indexing structure. Section 6 provides the experimental validation of the accuracy and efficiency of the proposed approach. Finally, section 7 concludes the chapter.

## 2. Background

Indexing of time-series has concentrated great attention in the research community, especially in the last decade, and this can partly be attributed to the explosion of database sizes. Characteristic examples are environmental data collected on a daily basis or satellite image databases of the earth[1]. If we would like to allow the user to explore these vast databases, we should organize the data in such a way, so that one can retrieve accurately and efficiently the data of interest.

More specifically, our objective is the automatic classification of time-series using Nearest Neighbor Classification (NNC). In NNC the time-series query is classified according to the majority of its nearest neighbors. The NNC is conceptually a simple technique, but provides very good results in practical situations. This classification technique is particularly suited for our setting, since we are going to use a *non-metric* distance function. Furthermore, the technique has good theoretical properties; it has been shown that the one nearest neighbor rule has asymptotic error rate that is at most twice the Bayes error rate [13].

So, the problem we consider is: "Given a database $\mathcal{D}$ of time-series and a query $\mathcal{Q}$ (not already in the database), find the sequence $\mathcal{T}$ that is closest to $\mathcal{Q}$." We need to define the following:

1. A realistic distance function that will match the user's perception of what is considered similar.

[1] `http://www.terraserver.com`

2. An efficient indexing scheme, which will speed up the user queries.

We will briefly discuss some issues associated with these two topics.

### 2.1. *Time Series Similarity Measures*

The simplest approach to define the distance between two sequences is to map each sequence into a vector and then use a p-norm to calculate their distance. The p-norm distance between two n-dimensional vectors $\vec{x}$ and $\vec{y}$ is defined as:

$$L_p(\vec{x}, \vec{y}) = \left( \sum_{i=1}^{n} |x_i - y_i|^p \right)^{\frac{1}{p}}$$

For $p = 2$ it is the well known Euclidean distance and for $p = 1$ the Manhattan distance.

Most of the related work on time-series has concentrated on the use of some metric $L_p$ Norm. The advantage of this simple model is that it allows efficient indexing by a dimensionality reduction technique [2,15,19,44]. On the other hand the model cannot deal well with outliers and is very sensitive to small distortions in the time axis. There are a number of interesting extensions to the above model to support various transformations such as scaling [10,36], shifting [10,21], normalization [21] and moving average [36]. Other recent works on indexing time series data for similarity queries assuming the Euclidean model include [25,26]. A domain independent framework for defining queries in terms of similarity of objects is presented in [24]. In [29], Lee *et al.* propose methods to index sequences of multidimensional points. The similarity model is based on the Euclidean distance and they extend the ideas presented by Faloutsos *et al.* in [16], by computing the distances between multidimensional Minimum Bounding Rectangles.

Another, more flexible way to describe similar but out-of-phase sequences can be achieved by using the *Dynamic Time Warping (DTW)* [38]. Berndt and Clifford [5] were the first that introduced this measure in the datamining community. Recent works on *DTW* are [27,31,45]. *DTW* has been first used to match signals in speech recognition [38]. *DTW* between two sequences $A$ and $B$ is described by the following equation:

$$DTW(A, B) = D_{base}(a_m, b_n) + \min\{DTW(\text{Head}(A), \text{Head}(B)), DTW(\text{Head}(A), B), DTW(A, \text{Head}(B))\}$$

where $D_{base}$ is some $L_p$ Norm and Head($A$) of a sequence $A$ are all the elements of $A$ except for $a_m$, the last one.

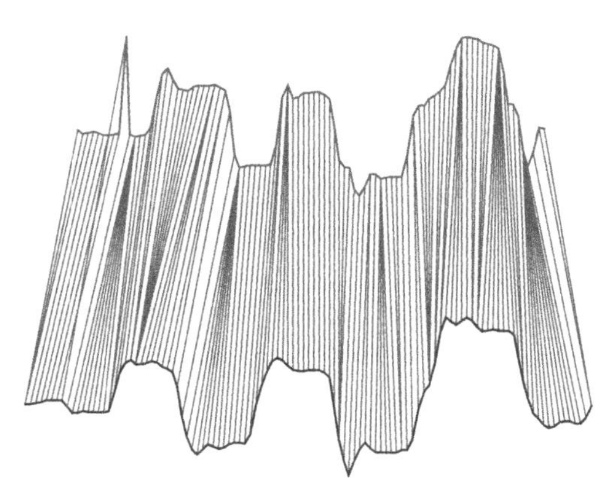

Fig. 2. The support of out-of-phase matching is important. However, the *DTW* matches all points (so, the outliers as well), therefore distorting the true distance between sequences. The *LCSS* model can support time-shifting and efficiently ignore the noisy parts.

The difficulties imposed by *DTW* include the fact that it is a non-metric distance function and also that its performance deteriorates in the presence of large amount of outliers (Figure 2). Although the flexibility provided by *DTW* is very important, nonetheless *DTW* is not the appropriate distance function for noisy data, since by matching all the points, it also matches the outliers distorting the true distance between the sequences.

Another technique to describe the similarity is to find the longest common subsequence *(LCSS)* of two sequences and then define the distance using the length of this subsequence [3,7,12,41]. The *LCSS* shows how well the two sequences can match one another if we are allowed to stretch them but we cannot rearrange the sequence of values. Since the values are real numbers, we typically allow approximate matching, rather than exact matching. A technique similar to our work is that of Agrawal *et al.* [3]. In this chapter the authors propose an algorithm for finding an approximation of the *LCSS* between two time-series, allowing local scaling (that is, different parts of each sequence can be scaled by a different factor before matching). In [7,12] fast probabilistic algorithms to compute the *LCSS* of two time series are presented.

Other techniques to define time series similarity are based on extracting certain features (Landmarks [32] or signatures [14]) from each time-series and then use these features to define the similarity. An interesting approach to represent a time series using the direction of the sequence at regular time intervals is presented in [35]. Ge and Smyth [18] present an alternative approach for sequence similarity that is based on probabilistic matching, using Hidden Markov Models. However the scalability properties of this approach have not been investigated yet. A recent work that proposes a method to cluster trajectory data is due to Gaffney and Smyth [17]. They use a variation of the EM (expectation maximization) algorithm to cluster small sets of trajectories. However, their method is a model based approach that usually has scalability problems.

The most related paper to our work is the Bozkaya *et al.* [8]. They discuss how to define similarity measures for sequences of multidimensional points using a restricted version of the edit distance which is equivalent to the $LCCS$. Also, they present two efficient methods to index the sequences for similarity retrieval. However, they focus on sequences of feature vectors extracted from images and they do not discuss transformations or approximate methods to compute the similarity.

Lately, there has been some work on indexing moving objects to answer spatial proximity queries (range and nearest neighbor queries) [1,28,39]. Also, in [33], Pfoser *et al.* present index methods to answer topological and navigational queries in a database that stores trajectories of moving objects. However these works do not consider a global similarity model between sequences but they concentrate on finding objects that are close to query locations during a time instant, or time period that is also specified by the query.

### 2.2. *Indexing Time Series*

Indexing refers to the organization of data in such a way, so as not only to group together similar items, but also to facilitate the pruning of irrelevant data. The second part is greatly affected from the fact whether our distance function is metric or not.

**Definition 1.** A distance function $d(x, y)$ between two objects $x$ and $y$ is metric when it complies to the following properties:

1. Positivity, $d(x, y) \geq 0$ and $d(x, y) = 0$ iff $x = y$.
2. Symmetry, $d(x, y) = d(y, x)$.
3. Triangle Inequality, $d(x, y) + d(y, z) \geq d(x, z)$.

**Example:** The Euclidean distance *is* a metric distance function. The pruning power of such a function can be shown in the following example(Figure 3); Assume we have a set of sequences, including sequences $Seq1$,

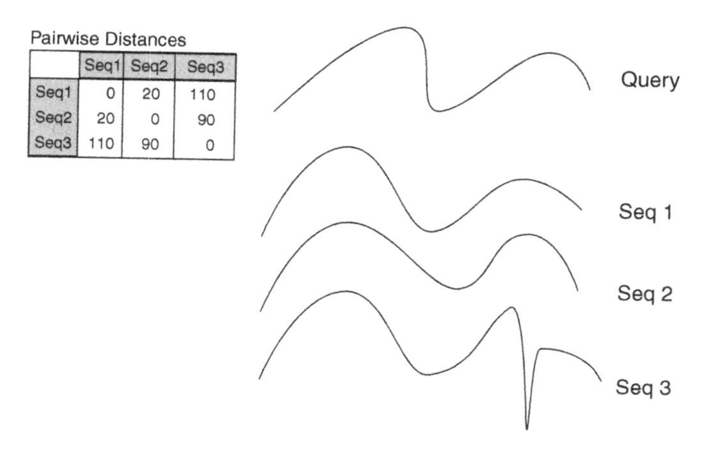
Pairwise Distances

| | Seq1 | Seq2 | Seq3 |
|---|---|---|---|
| Seq1 | 0 | 20 | 110 |
| Seq2 | 20 | 0 | 90 |
| Seq3 | 110 | 90 | 0 |



Fig. 3. Pruning power of triangle inequality.

*Seq*2, and *Seq*3, and we have already recorded their pair-wise distances in a table. Suppose we pose a query $Q$, and we want to find the closest sequence to $Q$. Assume we have already found a sequence, *best-match-so-far*, whose distance from $Q$ is 20. Comparing the query sequence $Q$ to *Seq2* yields $D(Q, Seq2) = 150$. However, because $D(Seq2,Seq1) = 20$ and it is true that: $D(Q,Seq1) \geq D(Q,Seq2) - D(Seq2,Seq1) \Rightarrow D(Q,Seq1) \geq 150 - 20 = 130$ we can safely prune *Seq1*, since it is not going to offer a better solution. Similarly, we can prune *Seq*3.

We are going to use non-metric distance functions. This choice imposes difficulties on how to prune data in an index, since we cannot apply the technique outlined above. As a result it is much more difficult to design efficient indexing techniques for non-metric distance functions. In the next section we show that despite this fundamental drawback it is important to use non-metric distance functions in conditions of noise. In section 5 we will also show that the proper choice of a distance measure can overcome this drawback to some degree.

## 2.3. *Motivation for Non-Metric Distance Functions*

Distance functions that are robust to extremely noisy data will typically violate the triangular inequality. These functions achieve this by not considering the most dissimilar parts of the objects. Moreover, they are useful, because they represent an accurate model of the human perception, since when comparing any kind of data (images, time-series etc), we mostly focus on the portions that are similar and we are willing to pay less attention to regions of great dissimilarity.

Non-metric distances are used nowadays in many domains, such as string (DNA) matching, collaborative filtering (where customers are matched with stored 'prototypical' customers) and retrieval of similar images from databases. Furthermore, psychology research suggests that human similarity judgments are also non-metric.

For this kind of data we need distance functions that can address the following issues:

- **Different Sampling Rates or different speeds.** The time-series that we obtain, are not guaranteed to be the outcome of sampling at fixed time intervals. The sensors collecting the data may fail for some period of time, leading to inconsistent sampling rates. Moreover, two time series moving at exactly the similar way, but one moving at twice the speed of the other will result (most probably) to a very large Euclidean distance.
- **Outliers.** Noise might be introduced due to anomaly in the sensor collecting the data or can be attributed to human 'failure' (e.g. jerky movement during a tracking process). In this case the Euclidean distance will completely fail and result to very large distance, even though this difference may be found in only a few points.
- **Different lengths.** Euclidean distance deals with time-series of equal length. In the case of different lengths we have to decide whether to truncate the longer series, or pad with zeros the shorter etc. In general its use gets complicated and the distance notion more vague.
- **Efficiency.** The similarity model has to be sufficiently complex to express the user's notion of similarity, yet simple enough to allow efficient computation of the similarity.

To cope with these challenges we use the Longest Common Subsequence (*LCSS*) model. The *LCSS* is a variation of the edit distance [30,37]. The basic idea is to match two sequences by allowing them to stretch, without rearranging the sequence of the elements but allowing some elements to be *unmatched.*

A simple extension of the *LCSS* model is not sufficient, because (for example) this model cannot deal with parallel movements. Therefore, we extend it in order to address similar problems. So, in our similarity model we consider a set of translations and we find the translation that yields the optimal solution to the *LCSS* problem.

## 3. Similarity Measures Based on *LCSS*

### 3.1. *Original Notion of* LCSS

Suppose that we have two time series $A = (a_1, a_2, \ldots, a_n)$ and $B = (b_1, b_2, \ldots, b_m)$. For $A$ let $\text{Head}(A) = (a_1, a_2, \ldots, a_{n-1})$. Similarly, for $B$.

**Definition 2.** The *LCSS* between $A$ and $B$ is defined as follows:

$$LCSS(A, B) = \begin{cases} 0 & \text{if } A \text{ or } B \text{ empty,} \\ 1 + LCSS(\text{Head}(A), B)) & \text{if } a_n = b_n, \\ \max(LCSS(\text{Head}(A), B), & \text{otherwise.} \\ \qquad LCSS(A, Head(B))) & \end{cases}$$

The above definition is recursive and would require exponential time to compute. However, there is a better solution that can be offered in $O(m^*n)$ time, using dynamic programming.

**Dynamic Programming Solution [11,42]**

The *LCSS* problem can easily be solved in quadratic time and space. The basic idea behind this solution lies in the fact that the problem of the sequence matching can be dissected in smaller problems, which can be combined after they are solved optimally. So, what we have to do is, solve a smaller instance of the problem (with fewer points) and then continue by adding new points to our sequence and modify accordingly the *LCSS*.

Now the solution can be found by solving the following equation using dynamic programming (Figure 4):

$$LCSS[i, j] = \begin{cases} 0 & \text{if } i = 0, \\ 0 & \text{if } j = 0, \\ 1 + LCSS[i-1, j-1] & \text{if } a_i = b_i, \\ \max(LCSS[i-1, j], LCSS[i, j-1]) & \text{otherwise.} \end{cases}$$

where $LCSS[i, j]$ denotes the longest common subsequence between the first $i$ elements of sequence $A$ and the first $j$ elements of sequence $B$. Finally, $LCSS[n, m]$ will give us the length of the longest common subsequence between the two sequences $A$ and $B$.

The same dynamic programming technique can be employed in order to find the Warping Distance between two sequences.

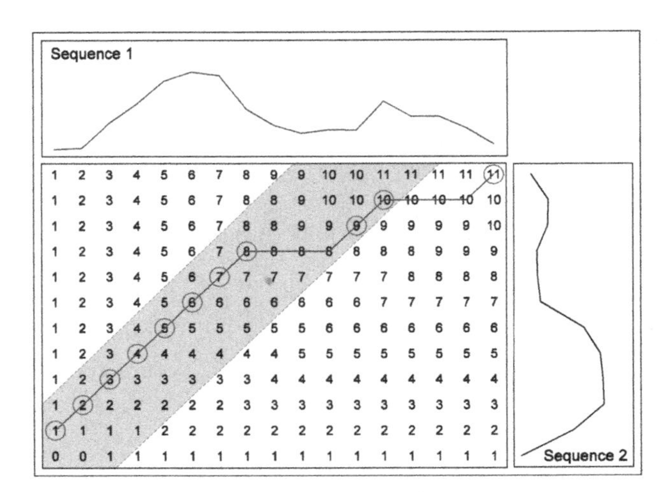


Fig. 4. Solving the *LCSS* problem using dynamic programming. The gray area indicates the elements that are examined if we confine our search window. The solution provided is still the same.

### 3.2. *Extending the* LCSS *Model*

Having seen that there exists an efficient way to compute the *LCSS* between two sequences, we extend this notion in order to define a new, more flexible, similarity measure. The *LCSS* model matches exact values, however in our model we want to allow more flexible matching between two sequences, when the values are within certain range. Moreover, in certain applications, the stretching that is being provided by the *LCSS* algorithm needs only to be within a certain range, too.

We assume that the measurements of the time-series are at fixed and discrete time intervals. If this is not the case then we can use interpolation [23,34].

**Definition 3.** Given an integer $\delta$ and a real positive number $\varepsilon$, we define the $LCSS_{\delta,\varepsilon}(A, B)$ as follows:

$$LCSS_{\delta,\varepsilon}(A,B) = \begin{cases} 0 \quad \text{if } A \text{ or } B \text{ is empty} \\ 1 + LCSS_{\delta,\varepsilon}(\text{Head}(A), \text{Head}(B)) \\ \qquad \text{if } |a_n - b_n| < \varepsilon \text{ and } |n - m| \le \delta \\ \max(LCSS_{\delta,\varepsilon}(\text{Head}(A), B), LCSS_{\delta,\varepsilon}(A, \text{Head}(B))) \\ \qquad \text{otherwise} \end{cases}$$

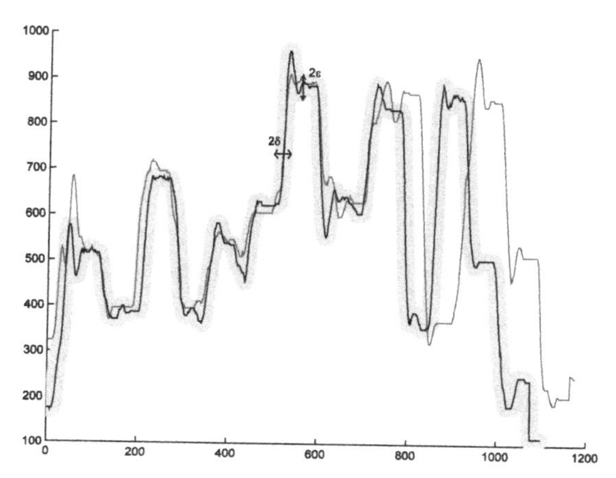


Fig. 5. The notion of the *LCSS* matching within a region of $\delta$ & $\varepsilon$ for a sequence. The points of the two sequences within the gray region can be matched by the extended *LCSS* function.

The constant $\delta$ controls how far in time we can go in order to match a given point from one sequence to a point in another sequence. The constant $\varepsilon$ is the matching threshold (see Figure 5).

The first similarity function is based on the *LCSS* and the idea is to allow time stretching. Then, objects that are close in space at different time instants can be matched if the time instants are also close.

**Definition 4.** We define the similarity function $S1$ between two sequences $A$ and $B$, given $\delta$ and $\varepsilon$, as follows:

$$S1(\delta, \varepsilon, A, B) = \frac{LCSS_{\delta,\varepsilon}(A, B)}{\min(n.m)}$$

Essentially, using this measure if there is a matching point within the region $\varepsilon$ we increase the *LCSS* by one.

We use function $S1$ to define another, more flexible, similarity measure. First, we consider the set of translations. A translation simply causes a vertical shift either up or down. Let $F$ be the family of translations. Then a function $f_c$ belongs to $F$ if $f_c(A) = (a_{x,1} + c, \ldots, a_{x,n} + c)$. Next, we define a second notion of the similarity based on the above family of functions.

**Definition 5.** Given $\delta, \varepsilon$ and the family $F$ of translations, we define the similarity function $S2$ between two sequences $A$ and $B$, as follows:

$$S2(\delta, \varepsilon, A, B) = \max_{fc \in F} S1(\delta, \varepsilon, A, f_c(B))$$

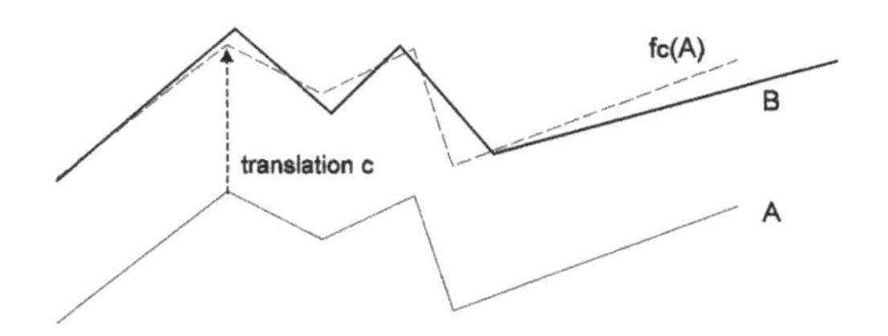


Fig. 6. Translation of sequence $A$.

So the similarity functions $S1$ and $S2$ range from 0 to 1. Therefore we can define the distance function between two sequences as follows:

**Definition 6.** Given $\delta, \varepsilon$ and two sequences $A$ and $B$ we define the following distance functions:

$$D1(\delta, \varepsilon, A, B) = 1 - S1(\delta, \varepsilon, A, B)$$

and

$$D2(\delta, \varepsilon, A, B) = 1 - S2(\delta, \varepsilon, A, B)$$

Note that $D1$ and $D2$ are *symmetric*. $LCSS_{\delta,\varepsilon}(A, B)$ is equal to $LCSS_{\delta,\varepsilon}(B, A)$ and the transformation that we use in $D2$ is translation which preserves the symmetric property.

By allowing translations, we can detect similarities between movements that are parallel, but not identical. In addition, the $LCSS$ model allows stretching and displacement in time, so we can detect similarities in movements that happen with different speeds, or at different times. In Figure 6 we show an example where a sequence $A$ matches another sequence $B$ after a translation is applied.

The similarity function $S2$ is a significant improvement over the $S1$, because: (i) now we can detect parallel movements, (ii) the use of *normalization* does not guarantee that we will get the best match between two time-series. Usually, because of the significant amount of noise, the average value and/or the standard deviation of the time-series that are being used in the normalization process can be distorted leading to improper translations.

### 3.3. *Differences between* DTW *and* LCSS

Time Warping and the $LCSS$ share many similarities. Here, we argue that the $LCSS$ is a better similarity function for correctly identifying noisy sequences and the reasons are:

1. Taking under consideration that a large portion of the sequences may be just outliers, we need a similarity function that will be robust under noisy conditions and will not match the incorrect parts. This property of the *LCSS* is depicted in the Figure 7. Time Warping by matching all elements is also going to try and match the outliers which, most likely, is going to distort the real distance between the examined sequences.

In Figure 8 we can see an example of a hierarchical clustering produced by the *DTW* and the *LCSS* distances between four time-series.

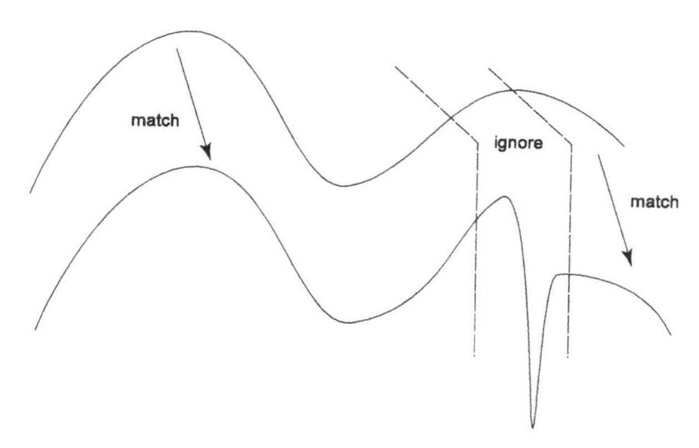


Fig. 7. Using the *LCSS* we only match the similar portions, avoiding the outliers.

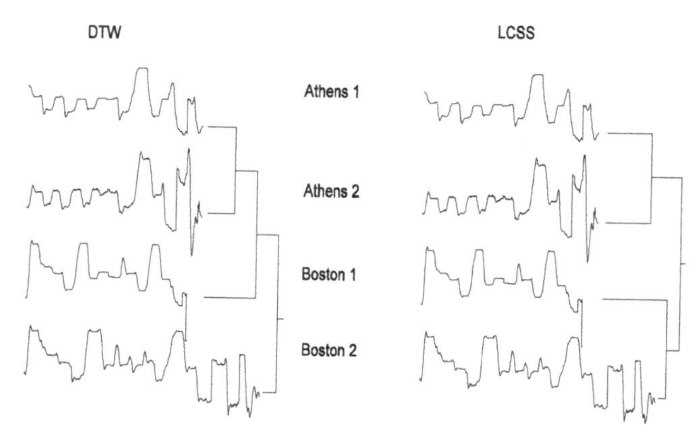


Fig. 8. Hierarchical clustering of time series with significant amount of outliers. *Left:* The presence of many outliers in the beginning and the end of the sequences leads to incorrect clustering. *DTW* is not robust under noisy conditions. *Right:* The *LCSS* focusing on the common parts achieves the correct clustering.

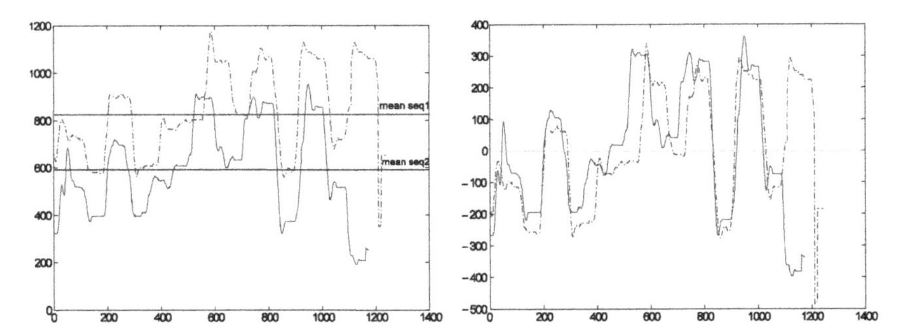


Fig. 9. *Left:* Two sequences and their mean values. *Right:* After normalization. Obviously an even better matching can be found for the two sequences.

The sequences represent data collected through a video tracking process (see Section 6). The *DTW* fails to distinguish the two classes of words, due to the great amount of outliers, especially in the beginning and in the end of the sequences. Using the Euclidean distance we obtain even worse results. Using the *LCSS* similarity measure we can obtain the most intuitive clustering as shown in the same figure. Even though the ending portion of the *Boston 2* time-series differs significantly from the *Boston 1* sequence, the *LCSS* correctly focuses on the start of the sequence, therefore producing the correct grouping of the four time-series.

2. Simply normalizing the time-series (by subtracting the average value) does not guarantee that we will achieve the best match (Figure 9). However, we are going to show in the following section, that we can try a set of translations which will provably give us the optimal matching (or close to optimal, within some user defined error bound).

## 4. Efficient Algorithms to Compute the Similarity

### 4.1. *Computing the Similarity Function S*1

To compute the similarity functions $S1$ we have to run a *LCSS* computation for the two sequences. The *LCSS* can be computed by a dynamic programming algorithm in $O(n^*m)$ time. However we only allow matchings when the difference in the indices is at most $\delta$, and this allows the use of a faster algorithm. The following lemma has been shown in [12].

**Lemma 1.** *Given two sequences A and B, with* $|A| = n$ *and* $|B| = m$, *we can find the* $LCSS_{\delta,\varepsilon}(A,B)$ *in* $O(\delta(n+m))$ *time.*

If $\delta$ is small, the dynamic programming algorithm is very efficient. However, for some applications $\delta$ may need to be large. In this case, we can speed-up the above computation using random sampling. Given two sequences $A$ and $B$, we compute two subsets $RA$ and $RB$ by sampling each sequence. Then we use the dynamic programming algorithm to compute the $LCSS$ on $RA$ and $RB$. We can show that, with high probability, the result of the algorithm over the samples, is a good approximation of the actual value. We describe this technique in detail in [40].

### 4.2. *Computing the Similarity Function S2*

We now consider the more complex similarity function $S2$. Here, given two sequences $A, B$, and constants $\delta, \varepsilon$, we have to find the translation $f_c$ that maximizes the length of the longest common subsequence of $A, f_c(B)$ ($LCSS_{\delta,\varepsilon}(A, f_c(B))$ over all possible translations.

A one dimensional translation $f_c$ is a function that adds a constant to all the elements of a 1-dimensional sequence: $f_c(x_1, \ldots, x_m) = (x_1 + c, \ldots, x_m + c)$.

Let the length of sequences $A$ and $B$ be $n$ and $m$ respectively. Let us also assume that the translation $f_{c1}$ is the translation that, when applied to $B$, gives a longest common subsequence $LCSS_{\delta,\varepsilon}(A, f_{c1}(B)) = a$, and it is also the translation that maximizes the length of the longest common subsequence:

$$LCSS_{\delta,\varepsilon}(A, f_{c1}(B)) = \max_{c \in R} LCSS_{\delta,\varepsilon}(A, f_c(B)).$$

The key observation is that, although there is an infinite number of translations that we can apply to $B$, each translation $f_c$ results to a longest common subsequence between $A$ and $f_c(B)$, and there is a finite set of possible longest common subsequences. In this section we show that we can efficiently enumerate a finite set of translations, such that this set provably includes a translation that maximizes the length of the longest common subsequence of $A$ and $f_c(B)$.

A translation by $c'$, applied to $B$ can be thought of as a linear transformation of the form $f(b_i) = b_i + c'$. Such a transformation will allow $b_i$ to be matched to all $a_j$ for which $|i - j| < \delta$, and $a_j - \varepsilon \leq f(b_i) \leq a_j + \varepsilon$.

It is instructive to view this as a stabbing problem: Consider the $O(\delta(n+m))$ vertical line segments $((b_i, a_j - \varepsilon), (b_i, a_j + \varepsilon))$, where $|i-j| < \delta$ (Figure 10).

These line segments are on a two dimensional plane, where on the $x$ axis we put elements of $B$ and on the $y$ axis we put elements of $A$. For every

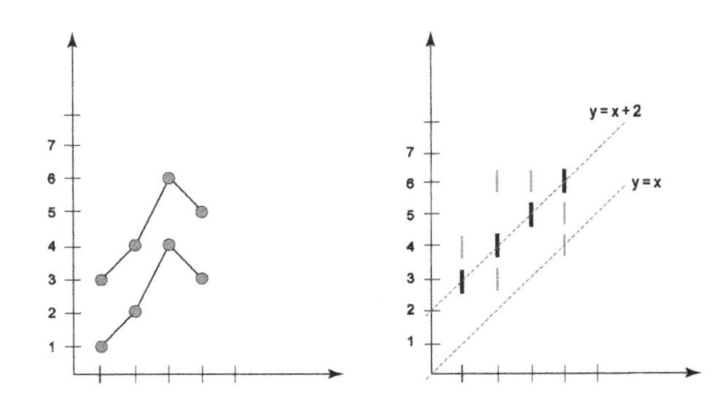


Fig. 10. An example of a translation. By applying the translation to a sequence and executing *LCSS* for $\delta = 1$ we can achieve perfect matching.

pair of elements $b_i, a_j$ in $A$ and $B$ that are within $\delta$ positions from each other (and therefore can be matched by the *LCSS* algorithm if their values are within $\varepsilon$), we create a vertical line segment that is centered at the point $(b_i, a_j)$ and extends $\varepsilon$ above and below this point. Since each element in $A$ can be matched with at most $2\delta + 1$ elements in $B$, the total number of such line segments is $O(\delta n)$.

A translation $f_{c'}$ in one dimension is a function of the form $f_{c'}(b_i) = b_i + c'$. Therefore, in the plane we described above, $f_{c'}(b_i)$ is a line of slope 1. After translating $B$ by $f_{c'}$, an element $b_i$ of $B$ can be matched to an element $a_j$ of $A$ if and only if the line $f_{c'}(x) = x + c'$ intersects the line segment $((b_i, a_j - \varepsilon), (b_i, a_j + \varepsilon))$.

Therefore each line of slope 1 defines a set of possible matchings between the elements of sequences $A$ and $B$. The number of intersected line segments is actually an upper bound on the length of the longest common subsequence because the ordering of the elements is ignored. However, two different translations can result to different longest common subsequences only if the respective lines intersect a different set of line segments. For example, the translations $f_0(x) = x$ and $f_2(x) = x+2$ in Figure 10 intersect different sets of line segments and result to longest common subsequences of different length.

The following lemma gives a bound on the number of possible different longest common subsequences by bounding the number of possible different sets of line segments that are intersected by lines of slope 1.

**Lemma 2.** *Given two one dimensional sequences $A$, $B$, there are $O(\delta(n + m))$ lines of slope 1 that intersect different sets of line segments.*

**Proof:** Let $f_{c'}(x) = x+c'$ be a line of slope 1. If we move this line slightly to the left or to the right, it still intersects the same number of line segments, unless we cross an endpoint of a line segment. In this case, the set of intersected line segments increases or decreases by one. There are $O(\delta(n+m))$ endpoints. A line of slope 1 that sweeps all the endpoints will therefore intersect at most $O(\delta(n+m))$ different sets of line segments during the sweep. □

In addition, we can enumerate the $O(\delta(n+m))$ translations that produce different sets of potential matchings by finding the lines of slope 1 that pass through the endpoints. Each such translation corresponds to a line $f_{c'}(x) = x + c'$. This set of $O(\delta(n+m))$ translations gives all possible matchings for a longest common subsequence of $A, B$. Since running the *LCSS* algorithm takes $O(\delta(n+m))$ we have shown the following theorem:

**Theorem 1.** *Given two sequences $A$ and $B$, with $|A| = n$ and $|B| = m$, we can compute the $S2(\delta, \varepsilon, A, B)$ in $O((n+m)^2\delta^2)$ time.*

### 4.3. *An Efficient Approximate Algorithm*

Theorem 1 gives an exact algorithm for computing $S2$, but this algorithm runs in quadratic time. In this section we present a much more efficient approximate algorithm. The key in our technique is that we can bound the difference between the sets of line segments that different lines of slope 1 intersect, based on how far apart the lines are.

Let us consider the $O(\delta(n+m))$ translations that result to different sets of intersected line segments. Each translation is a line of the form $f_{c'}(x) = x + c'$. Let us sort these translations by $c'$. For a given translation $f_{c'}$, let $L_{f_{c'}}$ be the set of line segments it intersects. The following lemma shows that neighbor translations in this order intersect similar sets of line segments.

**Lemma 3.** *Let $f_1(x) = x + c'_1, \ldots, f_N(x) = x + c'_N$ be the different translations for sequences $A_x$ and $B_x$, where $c'_1 \leq \cdots \leq c'_N$. Then the symmetric difference $L_{fi} \Delta L_{fj} \leq |i-j|$.*

**Proof:** Without loss of generality, assume $i < j$. The difference between $f_i$ and $f_{i+1}$ is simply that we cross one additional endpoint. This means that we have to either add or subtract one line segment to $L_{fi}$ to get $L_{fi+1}$. By repeating this step $(j-i)$ times we create $L_{fj}$. □

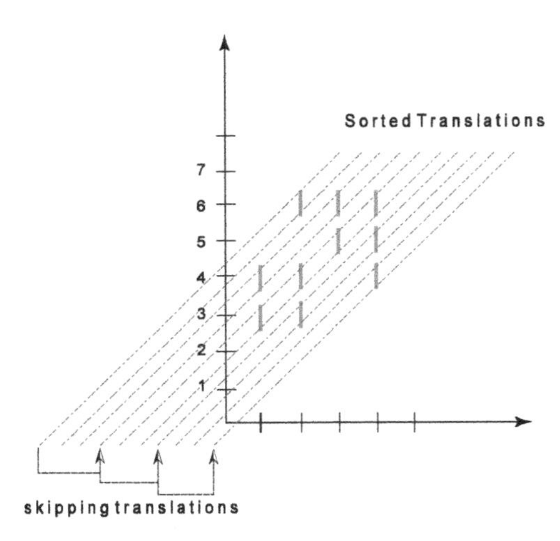


Fig. 11. If we can afford to be within a certain error range then we don't have to try *all* translations.

We can now prove our main theorem:

**Theorem 2.** *Given two sequences A and B, with* $|A| = n$ *and* $|B| = m$ *(suppose* $m < n$*), and a constant* $0 < \beta < 1$*, we can find an approximation* $AS2_{\delta,\beta}(A, B)$ *of the similarity* $S2(\delta, \varepsilon, A, B)$ *such that* $S2(\delta, \varepsilon, A, B) - AS2_{\delta,\beta}(A, B) < \beta$ *in* $O(n\delta^2/\beta)$ *time.*

**Proof:** Let $a = S2(\delta, \varepsilon, A, B)$. There exists a translation $f_i$ such that $L_{fi}$ is a superset of the matches in the optimal $LCSS$ of $A$ and $B$. In addition, by the previous lemma, there are $2b$ translations $(f_{i-b}, \ldots, f_{i+b})$ that have at most $b$ different matchings from the optimal.

Therefore, if we use the translations $f_{ib}$, for $i = 1, \ldots, \lceil \frac{\delta n}{b} \rceil$ in the ordering described above, we are within $b$ different matchings from the optimal matching of $A$ and $B$ (Figure 11). We can find these translations in $O(\delta n \log n)$ time if we find and sort all the translations.

Alternatively, we can find these translations in $O(\frac{\delta n}{b}\delta n)$ time if we run $\lceil \frac{\delta n}{b} \rceil$ quantile operations.

So we have a total of $\frac{\delta n}{b}$ translations and setting $b = \beta n$ completes the proof. □

Given sequences $A, B$ with lengths $n, m$ respectively ($m < n$), and constants $\delta$, $\beta$, $\varepsilon$, the approximation algorithm works as follows:

(i) *Find all sets of translations for sequences A and B.*

(ii) *Find the $i\beta n$-th quantiles for this set, $1 \leq i \leq \frac{2\delta}{\beta}$.*
(iii) *Run the $LCSS_{\delta,\varepsilon}$ algorithm on $A$ and $B$, for the computed $\frac{2\delta}{\beta}$ set of translations.*
(iv) *Return the highest result.*

## 5. Indexing Trajectories for Similarity Retrieval

In this section we show how to use the hierarchical tree of a clustering algorithm in order to efficiently answer nearest neighbor queries in a dataset of sequences.

The distance function $D2$ is not a metric because it does not obey the triangle inequality. This makes the use of traditional indexing techniques difficult. An example is shown in Figure 12, where we observe that $LCSS(\delta, \varepsilon, A, B) = 1$ and $LCSS(\delta, \varepsilon, B, C) = 1$, therefore the respective distances are both zero. However, obviously $D2(\delta, \varepsilon, A, C) > D2(\delta, \varepsilon, A, B) + D2(\delta, \varepsilon, B, C) = 0 + 0$.

We can however prove a weaker version of the triangle inequality, which can help us avoid examining a large portion of the database objects. First we define:

$$LCSS_{\delta,\varepsilon,F}(A, B) = \max_{fc \in F} LCSS_{\delta,\varepsilon}(A, f_c(B))$$

Clearly,

$$D2(\delta, \varepsilon, A, B) = 1 - \frac{LCSS_{\delta,\varepsilon,F}(\mathrm{A}, \mathrm{B})}{\min(|A|, |B|)}$$

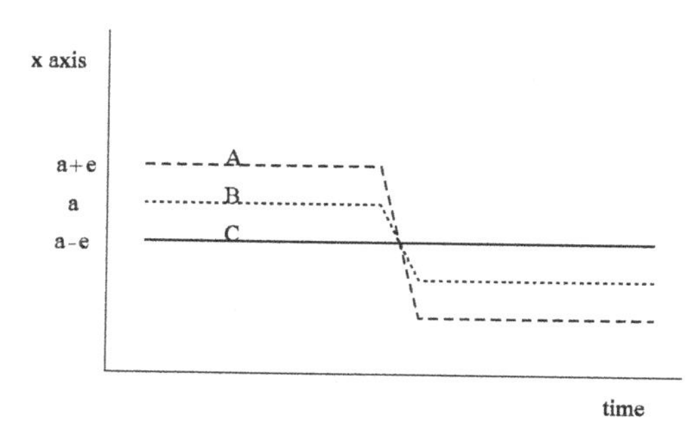


Fig. 12. An example where the triangle inequality does not hold for the new *LCSS* model.

(as before, $F$ is the set of translations). Now we can show the following lemma:

**Lemma 4.** *Given trajectories A,B,C:*

$$LCSS_{\delta,2\varepsilon}(A,C) > LCSS_{\delta,\varepsilon,F}(A,B) + LCSS_{\delta,\varepsilon,F}(B,C) - |B|$$

*where $|B|$ is the length of sequence B.*

**Proof:** Clearly, if an element of $A$ can match an element of $B$ within $\varepsilon$, and the same element of $B$ matches an element of $C$ within $\varepsilon$, then the element of $A$ can also match the element of $C$ within $2\varepsilon$. Since there are at least $|B| - (|B| - LCSS_{\delta,\varepsilon,F}(A,B)) - (|B| - LCSS_{\delta,\varepsilon,F}(B,C))$ elements of $B$ that match with elements of $A$ and with elements of $C$, it follows that $LCSS_{\delta,2\varepsilon,F}(A,C) > |B| - (|B| - LCSS_{\delta,\varepsilon,F}(A,B)) - (|B| - LCSS_{\delta,\varepsilon,F}(B,C)) = LCSS_{\delta,\varepsilon,F}(A,B) + LCSS_{\delta,\varepsilon,F}(B,C) - |B|$. □

### 5.1. *Indexing Structure*

We first partition all the sequences into sets according to length, so that the longest sequence in each set is at most $a$ times the shortest (typically we use $a = 2$.) We apply a hierarchical clustering algorithm on each set, and we use the tree that the algorithm produced as follows:

For every node $C$ of the tree we store the medoid ($M_C$) of each cluster. The medoid is the sequence that has the minimum distance (or maximum *LCSS*) from every other sequence in the cluster: $\max_{vi \in C} \min_{vj \in C} LCSS_{\delta,\varepsilon,F}(v_i, v_j, e)$. So given the tree and a query sequence $Q$, we want to examine whether to follow the subtree that is rooted at $C$. However, from the previous lemma we know that for any sequence $B$ in $C$:

$$LCSS_{\delta,\varepsilon,F}(B,Q) < |B| + LCSS_{\delta,2\varepsilon,F}(M_C,Q) - LCSS_{\delta,\varepsilon,F}(M_C,B)$$

or in terms of distance:

$$\begin{aligned} D2(\delta,\varepsilon,B,Q) &= 1 - \frac{LCSS_{\delta,\varepsilon,F}(B,Q)}{\min(|B|,|Q|)} \\ &> 1 - \frac{|B|}{\min(|B|,|Q|)} - \frac{LCSS_{\delta,2\varepsilon,F}(M_c,Q)}{\min(|B|,|Q|)} \\ &\quad + \frac{LCSS_{\delta,\varepsilon,F}(M_c,B)}{\min(|B|,|Q|)}. \end{aligned}$$

In order to provide a lower bound we have to maximize the expression $|B| - LCSS_{\delta,\varepsilon,F}(M_c, B)$. Therefore, for every node of the tree along with the medoid we have to keep the sequence $r_c$ that maximizes this expression. If the length of the query is smaller than the shortest length of the sequences we are currently considering we use that, otherwise we use the minimum and maximum lengths to obtain an approximate result.

## 5.2. *Searching the Index Tree for Nearest Trajectories*

We assume that we search an index tree that contains time-series with minimum length, $\min l$ and maximum length, $\max l$. For simplicity we discuss the algorithm for the 1-Nearest Neighbor query, where given a query sequence $Q$ we try to find the sequence in the set that is the most similar to $Q$. The search procedure takes as input a node $N$ in the tree, the query $Q$ and the distance to the closest time-series found so far. For each of the children $C$, we check if the child is a single sequence or a cluster. In case that it is a sequence, we just compare its distance to $Q$ with the current nearest sequence. If it is a cluster, we check the length of the query and we choose the appropriate value for $\min(|B|, |Q|)$. Then we compute a lower bound $L$ to the distance of the query with any sequence in the cluster and we compare the result with the distance of the current nearest neighbor *mindist*. We need to examine this cluster only if $L$ is smaller than *mindist*.

In our scheme we use an approximate algorithm to compute the $LCSS_{\delta,\varepsilon,F}$. Consequently, the value of $LCSS_{\delta,\varepsilon,F}(M_C, B)/\min(|B|, |Q|)$ that we compute can be up to $\beta$ times higher than the exact value. Therefore, since we use the approximate algorithm of section 3.2 for indexing trajectories, we have to subtract $\beta^* \min(|M_C|, |Q|)/\min(|B|, |Q|)$ from the bound we compute for $D2(\delta, \varepsilon, B, Q)$.

# 6. Experimental Evaluation

We implemented the proposed approximation and indexing techniques as they are described in the previous sections and here we present experimental results evaluating our techniques. We describe the datasets and then we continue by presenting the results. The purpose of our experiments is twofold: first, to evaluate the efficiency and accuracy of the approximation algorithm presented in Section 4 and second to evaluate the indexing technique that we discussed in the previous section. Our experiments were run on a PC Pentium III at 1 GHz with 128 MB RAM and 60 GB hard disk.

### 6.1. *Time and Accuracy Experiments*

Here we present the results of some experiments using the approximation algorithm to compute the similarity function $S2$. Our dataset here comes from marine mammals' satellite tracking data[2]. It consists of sequences of geographic locations of various marine animals (dolphins, sea lions, whales, etc) tracked over different periods of time, that range from one to three months (*SEALS* dataset). The length of the sequences is close to 100.

In Table 1 we show the computed similarity between a pair of sequences in the *SEALS* dataset. We run the exact and the approximate algorithm for different values of $\delta$ and $\varepsilon$ and we report here some indicative results. $K$ is the number of times the approximate algorithm invokes the *LCSS* procedure (that is, the number of translations $c$ that we try). As we can see using only a few translation we get very good results. We got similar results for synthetic datasets. Also, in Table 2 we report the running times to compute the similarity measure between two sequences of the same dataset. The approximation algorithm uses again from 15 to 60 different runs. The running time of the approximation algorithm is much faster even for $K = 60$.

As can be observed from the experimental results, the running times of the approximation algorithm is not proportional to the number of runs ($K$). This is achieved by reusing the results of previous translations and terminating early the execution of the current translation, if it is not going to yield a better result. The main conclusion of the above experiments is that the approximation algorithm can provide a very tractable time vs accuracy trade-off for computing the similarity between two sequences, when the similarity is defined using the *LCSS* model.

Table 1. Similarity values between two sequences from our SEALS dataset.

| $\delta$ | E | Similarity | | | | Error(%) for $K = 60$ |
|---|---|---|---|---|---|---|
| | | Exact | Approximate for $K$ tries | | | |
| | | | 15 | 30 | 60 | |
| 2 | 0.25 | 0.71134 | 0.65974 | 0.71134 | 0.696701 | 1.4639 |
| 2 | 0.5 | 0.907216 | 0.886598 | 0.893608 | 0.9 | 0.7216 |
| 4 | 0.25 | 0.71230 | 0.680619 | 0.700206 | 0.698041 | 1.2094 |
| 4 | 0.5 | 0.927835 | 0.908763 | 0.865979 | 0.922577 | 0.5258 |
| 6 | 0.25 | 0.72459 | 0.669072 | 0.698763 | 0.71134 | 1.325 |
| 6 | 0.5 | 0.938144 | 0.938 | 0.919794 | 0.92087 | 1.7274 |

[2]`http://whale.wheelock.edu/whalenet-stuff/stop_cover.html`

Table 2. Running time between two sequences from the SEALS dataset.

| $\delta$ | E | Running Time (sec) | | | | Error(%) for K = 60 |
|---|---|---|---|---|---|---|
| | | Exact | Approximate for K tries | | | |
| | | | 15 | 30 | 60 | |
| 2 | 0.25 | 0.06 | 0.0038 | 0.006 | 0.00911 | 6.586 |
| 2 | 0.5 | 0.04 | 0.0032 | 0.00441 | 0.00921 | 4.343 |
| 4 | 0.25 | 0.17 | 0.0581 | 0.00822 | 0.01482 | 11.470 |
| 4 | 0.5 | 0.1 | 0.00481 | 0.00641 | 0.01031 | 9.699 |
| 6 | 0.25 | 0.331 | 0.00731 | 0.01151 | 0.01933 | 17.123 |
| 6 | 0.5 | 0.301 | 0.0062 | 0.00921 | 0.01332 | 22.597 |

## 6.2. *Clustering using the Approximation Algorithm*

We compare the clustering performance of our method to the widely used Euclidean and *DTW* distance functions. Specifically:

(i) The Euclidean distance is only defined for sequences of the same length (and the lengths of our sequences vary considerably). We tried to offer the best possible comparison between every pair of sequences, by sliding the shorter of the two sequences across the longer one and recording their minimum distance. Like this we can get the best possible results out of the Euclidean method, imposing some time overhead. However, the computation of the Euclidean distance is the least complicated one.

(ii) The *DTW* can be used to match sequences of different length. In both *DTW* and Euclidean we normalized the data before computing the distances. Our method does not need any normalization, since it computes the necessary translations.

(iii) For *LCSS* we used a randomized version with and without sampling, and for various values of $\delta$. The time and the correct clusterings represent the average values of 15 runs of the experiment. This is necessary due to the randomized nature of our approach.

### 6.2.1. *Determining the Values for $\delta$ and $\varepsilon$*

The choice of values for the parameters $\delta$ and $\varepsilon$ are clearly dependent on the application and the dataset. For most datasets we had at our disposal we discovered that setting $\delta$ to more than 20–30% of the sequences length did not yield significant improvement. Furthermore, after some point the

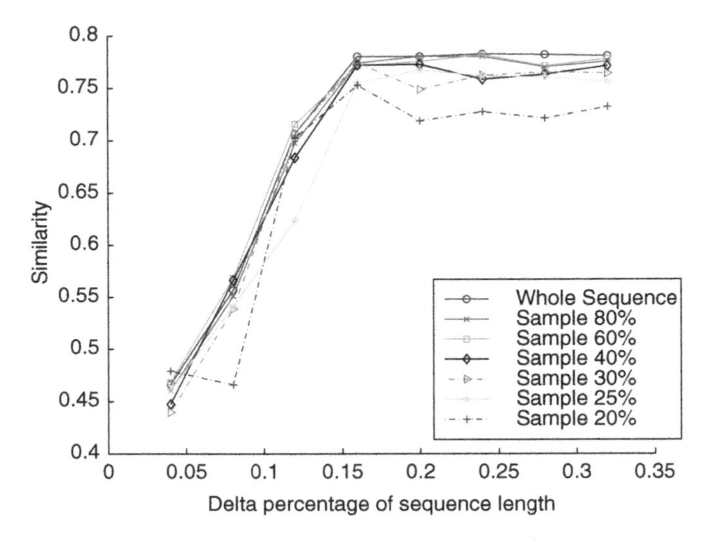


Fig. 13. In most datasets, the similarity stabilizes around a certain point after a value of $\delta$.

similarity stabilizes to a certain value (Figure 13). The determination of $\varepsilon$ is application dependent. In our experiments we used a value equal to some percentage (usually 0.3–0.5) of the smallest standard deviation between the two sequences that were examined at any time, which yielded good and intuitive results. Nevertheless, when we use the index the value of $\varepsilon$ has to be the same for all pairs of sequences.

If we want to provide more accurate results of the similarity between the two sequences, we can use weighted matching. This will allow us to give more gravity to the points that match very closely and less gravity to the points that match marginally within the search area of $\varepsilon$. Such matching functions can be found in Figure 14. If two sequence points match for very small $\varepsilon$, then we increase the *LCSS* by 1, otherwise we increase by some amount in the range $r$, where $0 \leq r < 1$.

### 6.2.2. *Experiment 1 — Video Tracking Data*

These time series represent the X and Y position of a human tracking feature (e.g. tip of finger). In conjunction with a *"spelling program"* the user can *"write"* various words [20]. In this experiment we used only the X coordinates and we kept 3 recordings for 5 different words. The data correspond to the following words: *'athens'*, *'berlin'*, *'london'*, *'boston'*, *'paris'*.

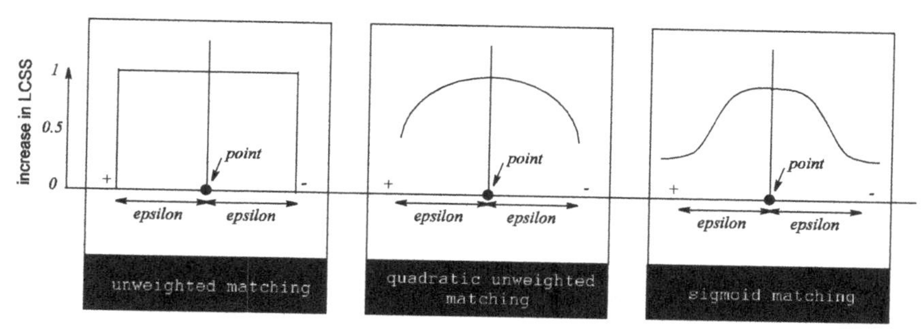


Fig. 14. Performing unweighted and weighted matching in order to provide more accuracy in the matching between sequences.

The average length of the series is around 1100 points. The shortest one is 834 points and the longest one 1719 points.

To determine the efficiency of each method we performed hierarchical clustering after computing the $N(N-1)/2$ pairwise distances for all three distance functions. We evaluate the total time required by each method, as well as the quality of the clustering, based on our knowledge of which word each sequence actually represents. We take all possible pairs of words (in this case $5 * 4/2 = 10$ pairs) and use the clustering algorithm to partition them into two classes. While at the lower levels of the dendrogram the clustering is subjective, the top level should provide an accurate division into two classes. We clustered using single, complete and average linkage. Since the best results for every distance function are produced using the *complete linkage*, we report only the results for this approach (Table 3, Fig. 15). The same experiment is conducted with the rest of the datasets. Experiments have been conducted for different sample sizes and values of $\delta$ (as a percentage of the original series length).

The results with the Euclidean and the Time Warping distance have many classification errors. For the *LCSS* the only real variations in the clustering are for sample sizes $s \leq 10\%$ (Figure 16). Still the average incorrect clusterings for these cases were constantly less than 2 (<1.85). For 15% sampling or more, there were no errors.

### 6.2.3. *Experiment 2 — Australian Sign Language Dataset (ASL)*[3]

The dataset consists of various parameters (such as the X ,Y, Z hand position, roll, pitch, yaw, thumb bend etc) tracked while different writers

[3]`http://kdd.ics.uci.edu`

Table 3. Results using the video tracking data some values of sample size s and $\delta$.

| Distance function | Time (sec) | Correct clusterings (out of 10) complete linkage |
|---|---|---|
| Euclidean | 31.28 | 2 |
| *DTW* | 227.162 | 5 |
| *S*2: | | |
| s-5%, $\delta$-25% | 2.645 | 8.45 |
| s-10%, $\delta$-25% | 8.240 | 9.85 |
| s-15%, $\delta$-25% | 15.789 | 10 |
| s-20%, $\delta$-25% | 28.113 | 10 |

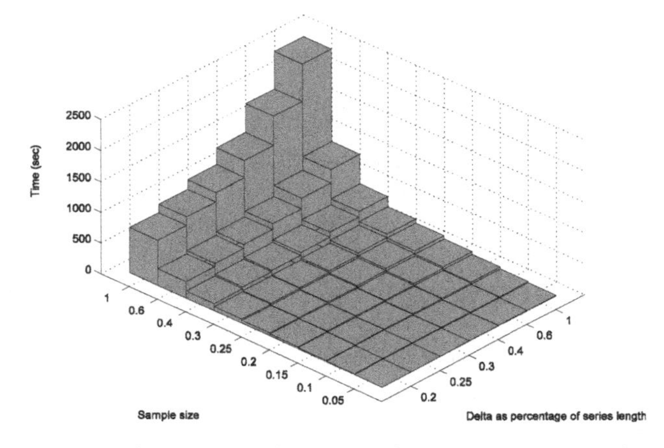


Fig. 15. Time required to compute the pairwise distance computations for various sample sizes and values of $\delta$.

sign one the 95 words of the ASL. These series are relatively short (50–100 points). We used only the X coordinates and collected 5 recordings of the following 10 words: *'Norway'*, *'cold'*, *'crazy'*, *'eat'*, *'forget'*, *'happy'*, *'innocent'*, *'later'*, *'lose'*, *'spend'*. This is the experiment conducted also in [27]. Examples of this dataset can be seen in Figure 17.

In this experiment all distance functions do not depict significant clustering accuracy, and the reason is that we used only 1 parameter out of the possible 10. Therefore accuracy is compromised. Specifically, the performance of the *LCSS* in this experiment is similar to the *DTW* (both recognized correctly 20 clusters). So both present equivalent accuracy levels, which is expected, since this dataset does not contain excessive noise and

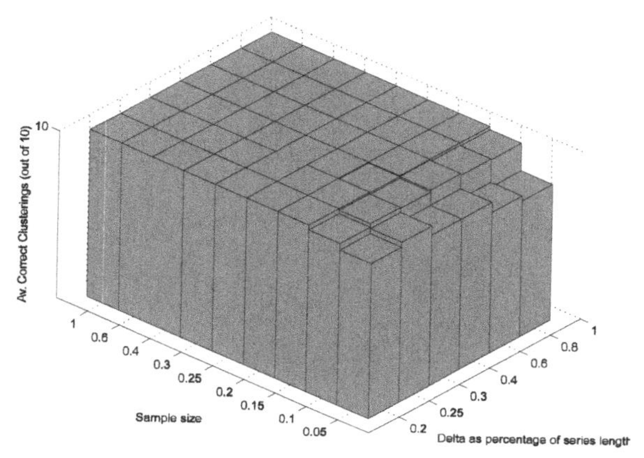


Fig. 16. Average number of correct clusterings (out of 10) for 15 runs of the algorithm using different sample sizes and $\delta$ values.

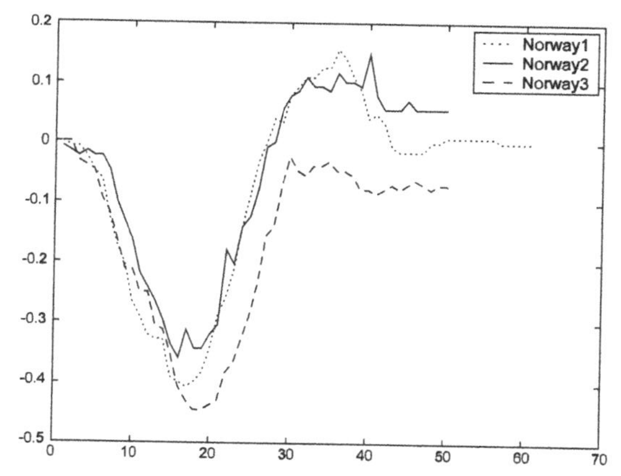


Fig. 17. Three recordings of the word *'norway'* in the Australian Sign Language. The graph depicts the $x$ position of the writer's hand.

furthermore the data seem to be already normalized and rescaled within the range $[-1, \ldots, 1]$. Therefore in this experiment we used also the similarity function $S1$ (no translation), since the translations were not going to achieve any further improvement (Table 4, Figure 18). Sampling is only performed down to 75% of the series length (these sequences are already short). As a consequence, even though we don't gain much in accuracy, our execution time is comparable to the Euclidean (without performing any translations).

Table 4. Results for ASL data and ASL with added noise.

| Distance function | Time (sec) | Correct clusterings (out of 45) ASL | Correct clusterings (out of 45) ASL with noise |
|---|---|---|---|
| Euclidean | 2.170 | 15 | 1 |
| *DTW* | 8.092 | 20 | 2 |
| *S*1: | | | |
| s-75%, $\delta$-40% | 1.200 | 15.450 | 9.933 |
| s-75%, $\delta$-100% | 2.297 | 19.231 | 10.4667 |
| s-100%, $\delta$-40% | 1.345 | 16.080 | 12.00 |
| s-100%, $\delta$-100% | 2.650 | 20.00 | 12.00 |

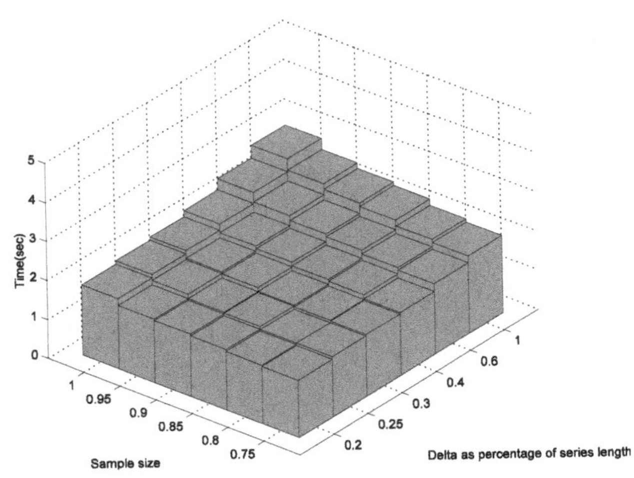


Fig. 18. *ASL* data: Time required to compute the pairwise distances of the 45 combinations (same for ASL and ASL with noise).

### 6.2.4. *Experiment 3 — ASL with Added Noise*

We added noise at every sequence of the ASL at a random starting point and for duration equal to the 15% of the series length. The noise was added using the function: $\langle \vec{x}_{\text{noise}}, \vec{y}_{\text{noise}} \rangle = \langle \vec{x}, \vec{y} \rangle + randn * rangeValues$, where *randn* produces a random number, chosen from a normal distribution with mean zero and variance one, and *rangeValues* is the range of values on X or Y coordinates. In this last experiment we wanted to see how the addition of noise would affect the performance of the three distance functions. Again, the running time is the same as with the original *ASL* data.

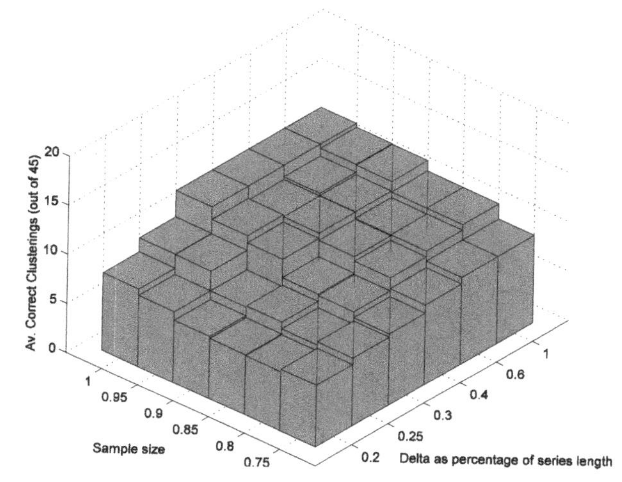


Fig. 19. Noisy *ASL* data: The correct clusterings of the *LCSS* method using complete linkage.

The *LCSS* proves to be more robust than the Euclidean and the *DTW* under noisy conditions (Table 4, Figures 17 and 18). The Euclidean again performed poorly, recognizing only 1 cluster, the *DTW* recognized 2 and the *LCSS* up to 12 clusters (consistently recognizing more than 6 clusters). These results are a strong indication that the models based on *LCSS* are generally more robust to noise.

## 6.3. *Evaluating the Quality and Efficiency of the Indexing Technique*

In this part of our experiments we evaluated the efficiency and effectiveness of the proposed indexing scheme. We performed tests over datasets of different sizes and different number of clusters. To generate large realistic datasets, we used real sequences (from the *SEALS* and *ASL* datasets) as "seeds" to create larger datasets that follow the same patterns. To perform tests, we used queries that do not have exact matches in the database, but on the other hand are similar to some of the existing sequences. For each experiment we run 100 different queries and we report the averaged results.

We have tested the index performance for different number of clusters in a dataset consisting of a total of 2000 sequences. We executed a set of $K$-Nearest Neighbor (K-NN) queries for $K$ 1, 5, 10, 15 and 20 and we plot the fraction of the dataset that has to be examined in order to guarantee

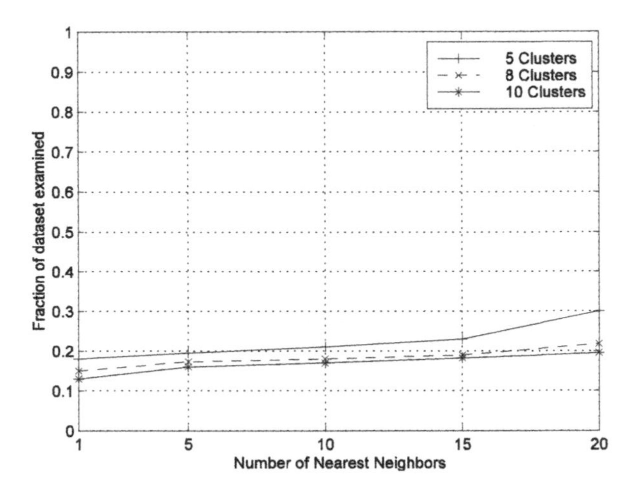


Fig. 20. Performance for increasing number of Nearest Neighbors.

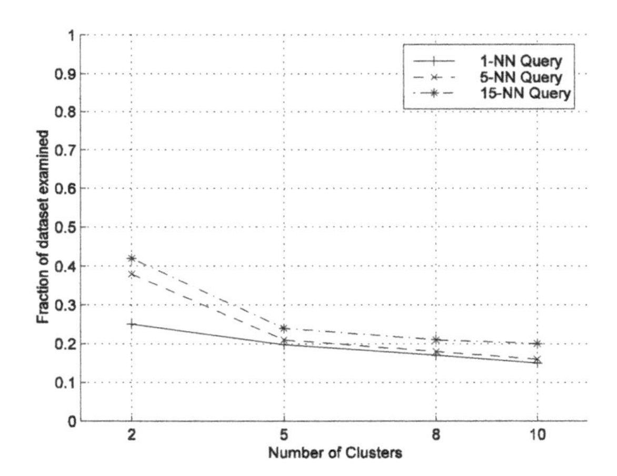


Fig. 21. Index performance for variable number of data clusters.

that we have found the best match for the K-NN query. Note that in this fraction we included the medoids that we check during the search since they are also part of the dataset.

In Figure 20 we show some results for $K$-Nearest Neighbor queries. We used datasets with 5, 8 and 10 clusters. As we can see the results indicate that the algorithm has good performance even for queries with large $K$. We also performed similar experiments where we varied the number of clusters in the datasets (Figure 21). As the number of clusters increased the performance of the algorithm improved considerably. This behavior is expected and it is similar to the behavior of recent proposed index structures for high dimensional data [6,9,22]. On the other hand if the dataset has no clusters, the performance of the algorithm degrades, since the majority of the sequences have almost the same distance to the query. This behavior follows again the same pattern of high dimensional indexing methods [6,43].

## 7. Conclusions

We have presented efficient techniques to accurately compute the similarity between time-series with significant amount of noise. Our distance measure is based on the *LCSS* model and performs very well for noisy signals. Since the exact computation is inefficient, we presented approximate algorithms with provable performance bounds. Moreover, we presented an efficient index structure, which is based on hierarchical clustering, for similarity (nearest neighbor) queries. The distance that we use is not a metric and therefore the triangle inequality does not hold. However, we prove that a similar inequality holds (although a weaker one) that allows to prune parts of the datasets without any false dismissals.

Our experiments indicate that the approximation algorithm can be used to get an accurate and fast estimation of the distance between two time-series even under noisy conditions. Also, results from the index evaluation show that we can achieve good speed-ups for searching similar sequences comparing with the brute force linear scan.

We plan to investigate biased sampling to improve the running time of the approximation algorithms, especially when full rigid transformations (e.g. shifting, scaling and rotation) are necessary. Another approach to index time-series for similarity retrieval is to use embeddings and map the set of sequences to points in a low dimensional Euclidean space [15]. The challenge of course is to find an embedding that approximately preserves the original pairwise distances and gives good approximate results to similarity queries.

## References


1. Agarwal, P.K., Arge, L., and Erickson, J. (2000). Indexing Moving Points. In *Proc. of the 19th ACM Symp. on Principles of Database Systems (PODS)*, pp. 175–186.
2. Agrawal, R., Faloutsos, C., and Swami, A. (1993). Efficient Similarity Search in Sequence Databases. In *Proc. of the 4th FODO*, pp. 69–84.

3. Agrawal, R. Lin, K. Sawhney, H.S., and Shim, K. (1995). Fast Similarity Search in the Presence of Noise, Scaling and Trans-Lation in Time-Series Databases. In *Proc. of (VLDB)*, pp. 490–501.
4. Barbara, D. (1999). Mobile Computing and Databases – A Survey. *IEEE TKDE*, pp. 108–117.
5. Berndt, D. and Clifford, J. (1994). Using Dynamic Time Warping to Find Patterns in Time Series. In *Proc. of KDD Workshop.*
6. Beyer, K., Goldstein, J., Ramakrishnan, R., and Shaft, U. (1999). When is "Nearest Neighbor" Meaningful? In *Proc. of ICDT, Jerusalem*, pp. 217–235.
7. Bollobas, B., Das, G., Gunopulos, D., and Mannila, H. (1997). Time-Series Similarity Problems and Well-Separated Geometric Sets. In *Proc. of the 13th SCG, Nice, France.*
8. Bozkaya, T., Yazdani, N., and Ozsoyoglu, M. (1997). Matching And Indexing Sequences of Different Lengths. In *Proc. of the CIKM, Las Vegas.*
9. Chakrabarti, K. and Mehrotra, S. (2000). Local Dimensionality Reduction: A New Approach to Indexing High Dimensional Spaces. In *Proc. of VLDB*, pp. 89–100.
10. Chu, K. and Wong, M. (1999). Fast Time-Series Searching with Scaling and Shifting. *ACM Principles of Database Systems*, pp. 237–248.
11. Cormen, T.H., Leiserson, C.E., and Rivest, R.L. (1990). *Introduction to Algorithms.* The MIT Electrical Engineering and Computer Science Series. MIT Press/McGraw Hill, 1990.
12. Das, G., Gunopulos, D. and Mannila, H. (1997). Finding Similar Time Series. In *Proc. of the First PKDD Symp.*, pp. 88–100.
13. Duda, R. and Hart, P. (1973). *Pattern Classification and Scene Analysis.* John Wiley and Sons, Inc.
14. Faloutsos, C., Jagadish, H., Mendelzon, A. and Milo, T. (1997). Signature Technique for Similarity-Based Queries. In *SEQUENCES 97.*
15. Faloutsos, C. and Lin, K.-I. (May, 1995). FastMap: A Fast Algorithm for Indexing, Datamining and Visualization of Traditional and Multimedia Datasets. In *Proc. ACM SIGMOD*, pp. 163–174.
16. Faloutsos, C., Ranganathan, M., and Manolopoulos, I. (1994). Fast Subsequence Matching in Time Series Databases. In *Proc. of ACM SIGMOD*, pp. 419–429.
17. Gaffney, S. and Smyth, P. (1999). Trajectory Clustering with Mixtures of Regression Models. In *Proc. of the 5th ACM SIGKDD, San Diego, CA*, pp. 63–72.
18. Ge, X. and Smyth, P. (2000). Deformable Markov Model Templates for Time-Series Pattern Matching. In *Proc. ACM SIGKDD.*
19. Gionis, A., Indyk, P., and Motwani, R. (1999). Similarity Search in High Dimensions Via Hashing. In *Proc. of 25th VLDB*, pp. 518–529.
20. Gips, J., Betke, M. and Fleming, P. (July, 2000). The Camera Mouse: Preliminary investigation of automated visual tracking for computer access. In *Proceedings of the Rehabilitation Engineering and Assistive Technology Society of North America 2000 Annual Conference*, pp. 98–100, Orlando, FL.

21. Goldin, D. and Kanellakis, P. (1995). On Similarity Queries for Time-Series Data. In *Proc. of CP '95, Cassis, France.*
22. Goldstein, J. and Ramakrishnan, R. (2000). Contrast Plots and p-Sphere Trees: Space vs. Time in Nearest Neighbour Searches. In *Proc. of the VLDB, Cairo*, pp. 429–440.
23. Grumbach, S., Rigaux, P., and Segoufin, L. (2000). Manipulating Interpolated Data is Easier than you Thought. In *Proc. of the 26th VLDB*, pp. 156–165.
24. Jagadish, H.V., Mendelzon, A.O., and Milo, T. (1995). Similarity-Based Queries. In *Proc. of the 14th ACMPODS*, pp. 36–45.
25. Kahveci, T. and Singh, A.K. (2001). Variable Length Queries for Time Series Data. In *Proc. of IEEE ICDE*, pp. 273–282.
26. Keogh, E., Chakrabarti, K., Mehrotra, S., and Pazzani, M. (2001). Locally Adaptive Dimensionality Reduction for Indexing Large Time Series Databases. In *Proc. of ACMSIGMOD*, pp. 151–162.
27. Keogh, E. and Pazzani, M. (2000). Scaling up Dynamic Time Warping for Datamining Applications. In *Proc. 6th Int. Conf. on Knowledge Discovery and Data Mining, Boston, MA.*
28. Kollios, G., Gunopulos, D., and Tsotras, V. (1999). On Indexing Mobile Objects. In *Proc. of the 18th ACM Symp. on Principles of Database Systems (PODS)*, pp. 261–272.
29. Lee, S.-L., Chun, S.-J., Kim, D.-H., Lee, J.-H., and Chung, C.-W. (2000). Similarity Search for Multidimensional Data Sequences. In *Proc. of ICDE*, pp. 599–608.
30. Levenshtein, V. (1966). Binary Codes Capable of Correcting Deletions, Insertions, and Reversals. In *Soviet Physics — Doklady 10, 10*, pp. 707–710.
31. Park, S., Chu, W., Yoon, J., and Hsu, C. (2000) Efficient Searches for Similar Subsequences of Different Lengths in Sequence Databases. In *Proc. of ICDE*, pp. 23–32.
32. Perng, S., Wang, H., Zhang, S., and Parker, D.S. (2000). Landmarks: A New Model for Similarity-Based Pattern Querying in Time Series Databases. In *Proc. of ICDE*, pp. 33–42.
33. Pfoser, D., Jensen, C., and Theodoridis, Y. (2000) Novel Approaches in Query Processing for Moving Objects. In *Proc. of VLDB, Cairo Egypt.*
34. Pfoser, D. and Jensen, C.S. (1999). Capturing the Uncertainty of Moving-Object Representations. *Lecture Notes in Computer Science*, 1651.
35. Qu, Y., Wang, C., and Wang, X. (1998). Supporting Fast Search in Time Series for Movement Patterns in Multiple Scales. In *Proc. of the ACM CIKM*, pp. 251–258.
36. Rafiei, D. and Mendelzon, A. (2000). Querying Time Series Data Based on Similarity. *IEEE Transactions on Knowledge and Data Engineering*, **12**(5), 675–693.
37. Rice, S.V., Bunke, H., and Nartker, T. (1997). Classes of Cost Functions for String Edit Distance. *Algorithmica*, **18**(2), 271–280.
38. Sakoe, H. and Chiba, S. (1978). Dynamic Programming Algorithm Optimization for Spoken Word Recognition. *IEEE Trans. Acoustics, Speech and Signal Processing (ASSP)*, **26**(1), 43–49.

39. Saltenis, S., Jensen, C., Leutenegger, S., and Lopez, M.A. (2000). Indexing the Positions of Continuously Moving Objects. In *Proc. of the ACM SIGMOD*, pp. 331–342.
40. Vlachos, M. (2001) Efficient Similarity Measures and Indexing Structures for Multidimensional Trajectories, *MSc Thesis*, UC Riverside.
41. Vlachos, M., Gunopulos, D., Kollios, G. (2002) Discovering Multidimensional Trajectories. In *Proc. of IEEE Int. Conference in Data Engineering (ICDE).*
42. Wagner, R.A. and Fisher, M.J. (1974). The String-To-String Correction Problem. *Journal of ACM*, **21**(1), 168–173.
43. Weber, R., Schek, H.-J., and Blott, S. (1998). A Quantitative Analysis and Performance Study for Similarity Search Methods in High-Dimensional Spaces. In *Proc. of the VLDB, NYC*, pp. 194–205.
44. Yi, B.-K. and Faloutsos, C. (2000). Fast Time Sequence Indexing for Arbitrary Lp Norms. In *Proc. of VLDB, Cairo Egypt.*
45. Yi, B.-K., Jagadish, H.V., and Faloutsos, C. (1998). Efficient Retrieval of Similar Time Sequences Under Time Warping. In *Proc. of the 14th ICDE*, pp. 201–208.

# CHAPTER 5

# CHANGE DETECTION IN CLASSIFICATION MODELS INDUCED FROM TIME SERIES DATA

Gil Zeira* and Oded Maimon†
*Department of Industrial Engineering*
*Tel-Aviv University, Tel-Aviv 69978, Israel*
E-mail: *zeiragil@post.tau.ac.il; †maimon@eng.tau.ac.il

Mark Last
*Department of Information Systems Engineering*
*Ben-Gurion University, Beer-Sheva 84105, Israel*
E-mail: mlast@bgumail.bgu.ac.il

Lior Rokach
*Department of Industrial Engineering*
*Tel-Aviv University, Tel-Aviv 69978, Israel*
E-mail: liorr@eng.tau.ac.il

Most classification methods are based on the assumption that the historic data involved in building and verifying the model is the best estimator of what will happen in the future. One important factor that must not be set aside is the time factor. As more data is accumulated into the problem domain, incrementally over time, one must examine whether the new data agrees with the previous datasets and make the relevant assumptions about the future. This work presents a new change detection methodology, with a set of statistical estimators. These changes can be detected independently of the data mining algorithm, which is used for constructing the corresponding model. By implementing the novel approach on a set of artificially generated datasets, all significant changes were detected in the relevant periods. Also, in the real-world datasets evaluation, the method produced similar results.

*Keywords*: Classification; incremental learning; time series; change detection; info-fuzzy network.

## 1. Introduction

As mass of data is incrementally accumulated into large databases over time, we tend to believe that the new data "acts" somehow resembling to the prior knowledge we have on the operation or facts that it describes. Change detection in time series is not a new subject and it has always been a topic of continued interest. For instance, Jones *et al.* [17] have developed a change detection model mechanism for serially correlated multivariate data. Yao [39] has estimated the number of change points in time series using the BIC criterion. However the change detection in classification is still not well elaborated.

There are many algorithms and methods that deal with the incremental learning problem, which is concerned with updating an induced model upon receiving new data. These methods are specific to the underlying data mining model. For example: Utgoff's method for incremental induction of decision trees (ITI) [35,36], Wei-Min Shen's semi-incremental learning method (CDL4) [34], David W. Cheung technique for updating association rules in large databases [5], Alfonso Gerevini's network constraints updating technique [12], Byoung-Tak Zhang's method for feedforwarding neural networks (SELF) [40], simple Backpropagation algorithm for neural networks [27], Liu and Setiono's incremental feature selection (LVI) [24] and more.

The main topic in most incremental learning theories is how the model (this could be a set of rules, a decision tree, neural networks, and so on) is refined or reconstructed efficiently as new amounts of data is encountered. This problem has been challenged by many of the algorithms mentioned above, and many of them performed significantly better than running the algorithm from scratch, generally when the records were received on-line and changes had a low magnitude. An important question that one must examine whenever a new mass of data is accumulated is "Is it really wise to keep on re-constructing or verifying the current model, when everything or something in our notion of the model could have significantly changed?" In other words, the main problem is not how to reconstruct better, but rather how to detect a change in a model based on a time series database.

Some researchers have proposed various representations of the problem of large time changing databases and populations including:

- Defining robustness and discovering robust knowledge from databases [15] or learning stable concepts [13] in domains with hidden changes in concepts.
- Identifying and modeling a persistent drift in a database [11].

- Adapting to Concept and Population Drift [14,18,19,21,34].
- Activity Monitoring [9].

Rather than challenging the problem of *detecting* significant changes, the above methods deal directly with data mining in changing environment.

This chapter introduces a novel methodology for detecting a significant change in a classification model of data mining, by identifying distinct categories of changes and implementing a set of statistical estimators. The major contribution of our change detection procedure (as will be described in later sections), is the ability to make a confident claim that the model that was pre-built based on a sufficiently large dataset is no longer valid for any future use such as prediction, rule induction, etc., and consequently, a new model must be constructed.

The rest of the chapter is organized as follows. In Section 2, we present our change detection procedure in data mining models, by defining the data mining classification model characteristics (2.1), describing the variety of possible significant changes (2.2), definition of the hypothesis testing (2.3), and the methodology for the change detection procedure (2.4). In Section 3, we describe an experimental evaluation of the change detection methodology by introducing artificial changes in databases and implementing the change detection methodology to identify these changes. Section 4 describes evaluation of the change detection methodology on real-world datasets. Section 5 presents validation of the work's assumptions and Section 6 concludes this chapter by summing-up the main contributions of our method, and presenting several options for future research in implementation and extension of our methodology.

## 2. Change Detection in Classification Models of Data Mining

### 2.1. *Classification Model Characteristics*

Classification is the task which involves constructing a model predicting the (usually categorical) label of a classifying attribute and using it to classify new data. The induced classification model can be represented as a set of rules (see the RISE system [8], the BMA method for rule induction [6], PARULEL and PARADISER etc.), a decision tree (see [34–36] ), neural networks (see [35]), information-theoretic connectionists networks (see IFN [25]) and so on.

Given a database $D$ which consist of $X$ set of records. The data mining model $M$ is one that by using an algorithm ($G$) generates a set of hypothesis tests $H$ within the available hypothesis space, which generalize the relationship between a set of candidate input variables and the target variable. The following notation is a general description of the data mining classification modeling task: Given a database $D$, containing a complete set of records $X = (A|T)$, where $A$ is a vector of candidate input variables (attributes from the examined phenomenon, which might have some influence on the target concept), and $T$ is a target variable (i.e. the target concept). Find the best set of hypothesis $H$ within the available hypothesis space, which generalize the relationship between a set of candidate input variables and the target variable (e.g. the model $M$) using some Data Mining algorithm. Each record is regarded as a complete set of conjunction between attributes and a target concept (or variable), such as

$$X_i = (A_{1i} = a_{1j}(1), A_{2i} = a_{2j}(2), \ldots, A_{ni} = a_{nj}(n), T_i = t_{j'}).$$

$A = (A_1, A_2 \ldots A_n)$ is a known set of attributes of the desired phenomenon and $T$ (also noted in $DM$ literature as $Y$) is a known discrete or continuous target variable or concept. The goal for the classification task is to generate best set of hypotheses to describe the model $M$ using an algorithm $G$. To simplify, this means, generating the following general equation:

$$M_G : A \neg \hat{T}. \tag{1}$$

It is obvious that not all attributes are proved to be statistically significant in order to be included in the set of hypothesizes.

In most algorithms, the database $D$ is divided into two parts — learning ($D_{\text{learn}}$) and validation ($D_{\text{val}}$) sets. The first is supposed to hold enough information for assembling a statistically significant and stable model based on the $DM$ algorithm. The second part is supposed to ensure that the algorithm performs its goals by validation of the built models on unseen records. Evaluating the prediction accuracy of any model $M$, which is built by a classification algorithm $G$, is commonly performed, by estimating the Validation Error Rate of the examined model.

When the database $D$ is not fixed but, alternatively, accumulated over time, the classification task should be altered: in every period $K$, a new set of records $X_K$ is accumulated to the database. $d_K$ will be the notation for the set of records $X_K$ that was added in the start of period $K$, and $D_K$ will be the notation for the accumulated database $D_k = \bigcup d_k$. Therefore, given a database $D_K$, containing a complete set of records $X_K$, generate the best set of hypothesis $H_K$ to describe the accumulated model $M_K$, and a new question is encountered: is $M_{K,G} = M_{K+1,G}$, in every $K = 1, \ldots, k$?

As noted before, most existing methods have dealt with "how can the model M be updated efficiently when a new period $K$ is encountered?" or "How can we adapt to the time factor?", rather than asking the following questions:

- "Was the model significantly changed during the period $K$?"
- "What was the nature of the change?"
- "Should we consider several of the past periods as redundant or not required in order for an algorithm $G$ to generate a better model $M$?"

Hence, the objective of this work is: define and evaluate a change detection methodology for identifying a significant change that happened during period $K$ in a classification model, which was incrementally built over periods 1 to $K-1$, based on the data that was accumulated during the period $K$.

## 2.2. *Variety of Changes*

There are various significant changes, which might occur when inducing the model $M$ using the algorithm $G$. There are several possible causes for significant changes in the data mining model:

(i) A change in the probability distribution of one or more of the input attributes ($A$). For example, if a database in periods 1 to $K-1$, consists of 45% males and 55% females, while in period $K$ all records represent males.

(ii) A change in the distribution of the target variable ($T$). For example, in the case of examining the rate of failures in a final exam based on the characteristics of the students in consecutive years. If in the year 1999 the average failure rate was 20% and in the year 2000 was 40%, then a change in the target distribution has occurred.

A change in the "patterns" (rules), which define the relationship of the input attributes to the target variable. That is a change in the model $M$, derived from a change in a set of hypothesis in $H$. For instance, in the case of examining the rate of failures in final exams based on the characteristics of the students in the course of consecutive years, if in years 1999 male students had 60% failures and female students had 5% failures, and in year 2000 the situation was the opposite, then it is obvious that there was a change in the patterns of behavior. This work defines this cause for a significant change in a Data Mining model $M$ by the following definition: A change C is encountered in the period $K$ if the validation error of the model $M_{K-1}$ (the model that is based on $D_{K-1}$) on the database $D_{K-1}$ is significantly different from the validation error rate of the model $M_{K-1}$ over $d_K$. $d_K$ consists of the accumulated set of records obtained in period $K$.

*Inconsistency of the DM algorithm.* The basic assumption is that the *DM* algorithms in use were proven to be consistent and therefore are not prone to be a major cause for a change in the model. Although, this option should not be omitted, this work does not intend to deal with consistency or inconsistency measures of existing *DM* algorithms. We just point out here that the *DM* algorithm that was chosen for our experiments (IFN) was shown to be stable in our previous work [21].

As noted in sub-section 2.1, the data mining model is generally described as: $M_G : A \neg \hat{T}$, that is a relationship between $A$ and $T$. The first cause is explained via a change in the set of attributes $A$, which can also cause a change in the target variable. For example, if the percent of women will rise from 50% to 80% and women drink more white wine than men, than the overall percent of white wine consumption ($T$) will also increase. This can also affect the overall error rate of the data mining model if most rules were generated for men. Also, if the cause of a change in the relevant period is a change in the target variable (e.g., France has stopped producing white wine, and the percent of white wine consumption has dropped from 50% to 30%), it is possible that the rules relating the relevant attributes to the target variable(s) will be affected.

Since it is not trivial to identify the dominant cause of change, this work claims that a variety of possible changes might occur in a relevant

Table 1. Definition of the variety of changes in a data mining model.

| Rules | $A$ | $T$ | Description |
|---|---|---|---|
| – | – | – | No change. |
| – | – | + | A change in the target variable. |
| – | + | – | A change in the attribute variable(s). |
| – | + | + | A change in the target and in the input variable(s). |
| + | – | – | A change in "patterns" (rules) of the data mining model. |
| + | – | + | A change in "patterns" (rules) of the data mining model, and a change in the target variable. |
| + | + | – | A change in "patterns" (rules) of the data mining model, and a change in the input variable(s). |
| + | + | + | A change in "patterns" (rules) of the data mining model, and a change in the target and the input variables. |

period. The change can be caused by one of the three causes that were mentioned above or a simple combination of them. Table 1 presents the possible combinations of significant causes in a given period.

The definition of the variety of possible changes in a data-mining model is quite a new concept. As noted, several researchers tended to deal with concept change, population change, activity monitoring, etc. The notion that all three major causes interact and affect each other is quite new and it is tested and validated in this work for the first time.

## 2.3. *Statistical Hypothesis Testing*

In order to determine whether or not a significant change has occurred in period $K$, a set of statistical estimators is presented in this chapter. The use of these estimators is subject to several conditions:

1. Every period contains a sufficient amount of training data in order to rebuild a model for that specific period. The decision of whether a period contains sufficient number of records should be based on the classification algorithm in use, the inherent noisiness of the training data, the acceptable difference between the training and validation error rates, and so on.
2. The same $DM$ algorithm is used in all periods to build the classification model (e.g., C4.5 or IFN).
3. The same method is used for estimating the validation error rate in all periods (e.g., 5-fold or 10-fold cross validation, 1/3 holdout, etc.).

*Detecting a change in "patterns" (rules).* "Patterns" (rules) define the relationship between input and target variables. Since massive data streams are usually involved in building an incremental model, we can safely assume that the true validation error rate of a given incremental model is accurately estimated by the previous $K-1$ periods. Therefore, a change in the rules ($R$) is encountered during period $K$ if the validation error of the model $M_{K-1}$ (the model that is based on $D_{K-1}$) on the database $D_{K-1}$ is significantly different from the validation error rate of the model $M_{K-1}$ over $d_K$ (the records obtained during period $K$).

Therefore, the parameter of interest for the statistical hypothesis testing is the true validation error rate and the null hypothesis is as follows:

$$H_0 : \hat{e}_{M_{K-1},K} = \hat{e}_{M_{K-1},K-1},$$
$$H_1 : \hat{e}_{M_{K-1},K} \neq \hat{e}_{M_{K-1},K-1},$$

where:

$\hat{e}_{M_{K-1},K-1}$ is the validation error rate of model $M_{K-1}$ measured on $D_{K-1}$ set of records (the standard validation error of the model).

$\hat{e}_{M_{K-1},K}$ is the validation error rate of the aggregated model $M_{K-1}$ on the set of records $d_K$.

In order to detect a significant difference between the two error rates it is needed to test the following statistic (two sided hypothesis):

$$|\hat{d}| = |\hat{e}_{M_{K-1},K} - \hat{e}_{M_{K-1},K-1}|,$$

$$\hat{\sigma}_d^2 = \frac{\hat{e}_{M_{K-1},K} - (1 - \hat{e}_{M_{K-1},K})}{n_K} + \frac{\hat{e}_{M_{K-1},K-1}(1 - \hat{e}_{M_{K-1},K-1})}{n_{K-1(\text{val})}},$$

If $|\hat{d}| \geq z_{(1-\alpha/2)} \cdot \sqrt{\hat{\sigma}_d^2}$, then reject $H_0$. A change has occured in period K. $n_{K-1(\text{val})} = |D_{K-1(\text{val})}|$ is the number of records which were selected for validation from periods $1, \ldots, K-1$ and $n_K = |d_K|$ is the number of records in period $K$.

The foundations for the above way of hypothesis testing when comparing error rates of classification algorithms can be found in [29].

*Detecting changes in variable distributions.* The second statistical test is Pearson's chi-square statistic for comparing multinomial variables (see [28]). This test examines whether a sample of the variable distribution is drawn from the matching probability distribution known as the true distribution of that variable. The objective of this estimator in the change detection procedure is to validate our assumption that the distribution of an input or a target variable has significantly changed in statistical sense. Again, since massive data streams are usually involved in building an incremental model, it is safe to assume that the stationary distribution of any variable in a given incremental model can be accurately estimated by the previous $K-1$ periods.

The following null hypothesis is tested for every variable of interest $X$:

$H_0$: the variable $X$'s distribution is stationary (time-invariant).

$H_1$: otherwise.

The decision is based on the following formula:

$$X_p^2 = n_K \cdot \sum_{i=1}^{j} \frac{(x_{iK}/n_K - x_{iK-1}/n_{K-1})^2}{x_{iK-1}/n_{K-1}}, \tag{2}$$

where:

$n_K$ is the number of records in the $K$th period.

$n_{K-1}$ is the number of records in periods $1, \ldots, K-1$.

$x_{iK}$ is the number of records in the $i$th class of variable $X$ in the $K$ period.

$x_{iK-1}$ is the number of records in the $i$th class of variable $X$ in periods $1, \ldots, K-1$.

If $X_p^2 > \chi_{1-\alpha}^2(j-1)$, where $j$ is the number of classes of the tested variable, then the null hypothesis that the variable $X$'s distribution has been stationary in period $K$ like in the previous periods is rejected.

The explanation of Pearson's statistical hypothesis testing is provided in [28].

## 2.4. *Methodology*

This section describes the algorithmic usage of the previous estimators:

**Inputs:**

- $G$ is the $DM$ algorithm used for constructing the classification model (e.g., C4.5 or IFN).
- $M$ is the classification model constructed by the $DM$ algorithm (e.g., a decision tree).
- $V$ is the validation method in use (e.g., 5-fold cross-validation).
- $K$ is the cumulative number of periods in a data stream.
- $\alpha$ is the desired significance level for the change detection procedure (the probability of a false alarm when no actual change is present).

**Outputs:**

- $CD(\alpha)$ is the error-based change detection estimator (1 – $p$-value).
- $XP(\alpha)$ is the Pearson's chi-square estimator of distribution change (1 – $p$-value).

## 2.5. *Change Detection Procedure*

*Stage 1:*

For periods $K-1$ build the model $M_{K-1}$ using the $DM$ algorithm $G$.

Define the data set $D_{K-1(\mathrm{val})}$.

Count the number of records $n_{K-1} = |D_{K-1(\mathrm{val})}|$.

Calculate the validation error rate $\hat{e}_{M_{K-1},K-1}$ according to the validation method $V$.

Calculate $x_{iK-1}$, $n_{K-1}$ for every input and target variable existing in periods $1, \ldots, K-1$.

*Stage 2:*

For period $K$, define the set of records $d_K$.
Count the number of records $n_K = |d_K|$.
Calculate $\hat{e}_{M_{K-1},K}$ according to the validation method $V$.
Calculate the difference

$$d = ABS(\hat{e}_{M_{K-1},K} - \hat{e}_{M_{K-1},K-1}), \quad \hat{\sigma}_d^2, \quad H_0 = z_{(1-\alpha/2)} \cdot \sqrt{\hat{\sigma}_d^2}.$$

Calculate and Return $CD(\alpha)$.

*Stage 3:*

For every input and target variable existing in periods $1, \ldots, K$:
Calculate: $x_{iK}$, $n_K$ and $X_p^2$.
Calculate and Return $XP(\alpha)$.

It is obvious that the complexity of this procedure is at most $O(n_K)$. Also, it is very easy to store information about the distributions of target and input variables in order to simplify the change detection methodology.

Based on the outputs of the change detection procedure, the user can make a distinction between the eight possible variations of a change in the data mining classification model (see sub-section above). Knowing the causes of the change (if any), the user of this new methodology can act in several ways, including reapplying the algorithm from scratch to the new data, absorb the new period and update the model by using an incremental algorithm, make $K' = K + 1$ and perform the change detection procedure again for the next period, explore the type of the change and its magnitude and effect on other characteristics of the $DM$ model, and incorporate other known methods dealing with the specific change(s) detected. One may also apply multiple model approaches such as boosting, voting, bagging, etc.

The methodology is not restricted to databases with a constant number of variables. The basic assumption is that if the addition of a new variable will influence the relationship between the target variable and the input variables in a manner that changes the validation accuracy, it will be identified as a significant change.

The procedure has three major stages. The first one is designed to perform initialization of the procedure. The second stage is aimed at detecting a significant change in the "patterns" (rules) of the pre-built data-mining model, as described in the previous section. The third stage is designated to test whether the distribution of one or more variable(s) in the set of input or target variable(s) has changed.

The basic assumption for using this procedure is the use of sufficient statistics for a run of the algorithm in every period. As indicated above, if this assumption is not valid, it is necessary to merge two or more periods to maintain statistically significant outcomes.

## 3. Experimental Evaluation

### 3.1. *Design of Experiments*

In order to evaluate the change detection algorithm, a set of artificially generated datasets were built based on the following characteristics:

- Pre-determined definition and distribution of all variables (candidate input and target).
- Pre-determined set of rules.
- Pure random generation of records.
- Non-correlated datasets (between periods).
- Minimal randomly generated noise.
- No missing data.

In all generated datasets, we have introduced and tested a series of artificially non-correlated changes of various types.

All datasets were mined with the IFN (Information-Fuzzy Network) program (version 1.2 beta), based on the Information-Theoretic Fuzzy Approach to Knowledge Discovery in Databases [Maimon and Last (2000)]. This novel method, developed by Mark Last and Oded Maimon was shown to have better dimensionality reduction capability, interpretability, and stability than other data mining methods [e.g., see Last *et al.* (2002)] and was therefore found suitable for this study.

This chapter uses two sets of experiments to evaluate the performance of the change detection procedure:

- The first set is aimed to estimate the hit rate (also called the "true positive rate") of the change detection methodology. Twenty four different changes in two different databases were designed under the rules mentioned above in order to confirm the expected outcomes of the change detection procedure. Table 2 below summarizes the distribution of the artificially generated changes in experiments on Database#1 and Database#2.
- All changes were tested independently under the minimum 5% confidence level by the following set of hypothesis. All hypotheses were tested separately with the purpose of evaluating the relationship of all tests.

- Also, 12 non-correlated sets of experiments, which do not contain any change, were implemented in order to estimate the actual 1st type error rate ($\alpha$) — the "detection" of a change that doesn't occur. This error is also called "false positive rate" or "false alarm rate".

The expected outcomes of the hypothesis testing on the various types of changes are described in the following tables.

It is expected that every uncorrelated change implemented will lead to the associated outcome of the relevant test as indicated in Tables 3 and 4.

The second part of the experiments on artificially generated datasets evaluated the change detection procedure on time series data. During the third and sixth periods (out of seven consecutive periods) two changes (in classification "rules") were introduced into the database and implemented disregarding the previous and the consecutive periods. It is expected that

Table 2. Distribution of the artificially generated changes in experiment part1.

| | Change in $A$ | Change in $Y$ | Change in the "Rules" | Total |
|---|---|---|---|---|
| Database #1 | 4 | 2 | 6 | 12 |
| Database #2 | 4 | 2 | 6 | 12 |
| Sum | 8 | 4 | 12 | 24 |

Table 3. Change Detection in Rules (CD).

| | Change in $A$ | Change in $T$ | Change in the "patterns" (rules) |
|---|---|---|---|
| CD (5%) | TRUE/FALSE (N/A) | TRUE/FALSE (N/A) | TRUE |

Table 4. Pearson's estimator for comparing distributions of variables (XP).

| Database section tested | Change in $A$ | Change in $T$ | Change in the "patterns" (rules) |
|---|---|---|---|
| **Candidate** | | | |
| Variables XP(5%) | TRUE | TRUE/FALSE (N/A) | TRUE/FALSE (N/A) |
| **Target** | | | |
| Variables XP(5%) | TRUE/FALSE (N/A) | TRUE | TRUE/FALSE (N/A) |
| **Candidate & Target** | | | |
| Variables XP(5%) | TRUE | TRUE | TRUE/FALSE (N/A) |

the Change Detection Procedure will reveal the changes in these periods only and not in any other period.

### 3.2. *Results — Part 1 (Hit Rate and False Alarm Rate)*

Based on the 24 change detection trials the following accumulated results have been obtained, as described thoroughly in Tables 5 and 6:

- All artificial changes in the "patterns" (rules) relating the input variables to the target variable were detected by the CD (5%) procedure. The average detection rate of CD was 100% meaning that all changes were recognized as significant by the procedure.
- Two trials generated by an artificial change in the target variable were detected by the CD procedure. The average detection rate of CD in these trials was 100%.
- According to the XP hypothesis testing, 50% of changes in the candidate input variables produced a change in the target variable, resulting in an average detection rate of 100% for input attributes and 98% for the target attribute.
- All the changes, which were mainly introduced to affect the target variable, were recognized as significant by the XP procedure.
- All the changes, which were mainly introduced to affect the relationship between the target and the candidate input variables, were recognized as significant by the XP procedure, with an average of 100% detection rate.
- The actual 2nd type error (false negative) rate of the change detection methodology in this experiment is 0%. No change was left undetected by the methodology.
- The actual 1st type error (false alarm) rate of the change detection methodology in these experiments is up to 6%. When implementing the methodology in trials which do not include changes, the CD estimator did not produce false alarms (0% error rate) and the XP estimator failed to produce accurate estimation in only five out of 84 tests (5.9%).

To conclude this part of the experiments with various possible changes in the data mining model, the following assumptions are validated by the results which were described above: a major change can cause secondary effects and generate other changes. The Change Detection procedure can detect these changes with theoretically and actually low Type 1 and Type 2 error rates.

Table 5. Average detection rate of the 24 trials (all significant attributes included in XP).

| Type of change: | $A$ | $T$ | $R$ | Average |
|---|---|---|---|---|
| Number of changes: | 8 | 4 | 12 | 24 |
| CD | 67% | 99% | 100% | 90% |
| XP(candidate) | 100% | | | 100% |
| XP(target) | | 100% | 100% | 100% |
| XP(target+candidate) | 99% | 98% | 98% | 99% |

Table 6. Distribution of 5% significance outcomes in the 24 trials.

| Type of change: | $A$ | $T$ | $R$ | sum |
|---|---|---|---|---|
| Number of changes: | 8 | 4 | 12 | 24 |
| CD | 0 | 3 | 12 | 15 |
| XP(candidate) | 4 | | | 4 |
| XP(target) | | 1 | 9 | 10 |
| XP(target+candidate) | 4 | 3 | 3 | 10 |

### 3.3. *Results — Part 2 (Time Series Data)*

Table 7 and Figure 1 describe the outcomes of applying the IFN algorithm to seven consecutive periods from the same artificially generated database (Database#1), with two changes introduced, and using the change detection methodology to detect significant changes that have occurred during these periods.

The results of the experiment can be summarized as follows:

- All changes are detected by the change detection procedure only in the relevant period where the corresponding change has occurred.
- Whenever CD attained a significant level (above 5%), XP also reached a significant value (more than 5%).
- The effect of the change decreases after one period for CD (when the change occurs in period $K$, the significance level in period $K+1$ is most similar to period $K-1$). Hence, after one period, if a change is not properly detected, it will be absorbed and discarded.
- The XP estimator of distribution change in input and target variables is quite sensitive to the effect of any change.

These results support the assumptions that a change in the target variable does not necessarily affect the classification "rules" of the database and that a change can mainly be detected in the first period after its occurrence.

Table 7. Results of the CD hypothesis testing on an artificially generated time series database.

| Period | Change introduced | CD | | | | | XP |
|---|---|---|---|---|---|---|---|
| | | $e_{MK-1,K}$ | $e_{MK-1,K-1}$ | $d$ | $H(95\%)$ | $1 - p$-value | $1 - p$-value |
| 1 | No | — | — | — | — | — | — |
| 2 | No | 20.30% | 24.00% | 3.70% | 4.30% | 91.90% | 92.50% |
| **3** | **Yes** | **32.80%** | **19.80%** | **13.00%** | **3.20%** | **100.00%** | **100.00%** |
| 4 | No | 26.60% | 24.60% | 2.00% | 2.80% | 88.20% | 100.00% |
| 5 | No | 26.30% | 26.40% | 0.10% | 2.50% | 52.60% | 99.90% |
| **6** | **Yes** | **18.80%** | **27.20%** | **8.40%** | **2.20%** | **100.00%** | **100.00%** |
| 7 | No | 22.10% | 22.00% | 0.10% | 2.00% | 53.40% | 52.80% |

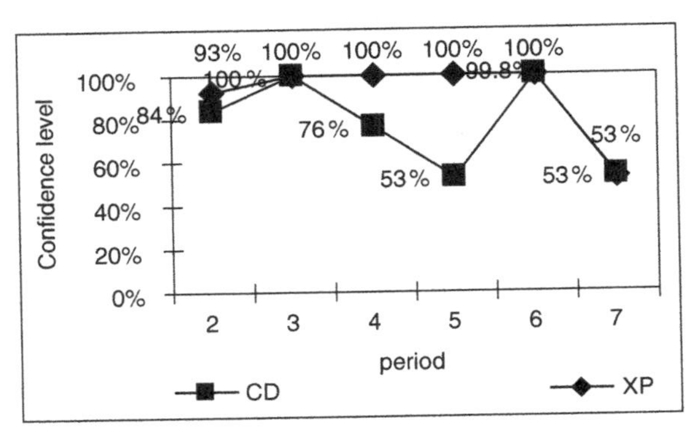


Fig. 1. Summary of implementing the change detection methodology on an artificially generated time series database ($1 - p$-value).

Table 8. Influence of discarding the detected change (Illustration).

| Candidate Variable | | Prediction | | Match |
|---|---|---|---|---|
| Products | Holidays | Periods 1–5 | Period 6 | |
| 0 | * | 1 | 1 | Yes |
| 1 | * | 1 | 1 | Yes |
| 2 | * | 1 | 1 | Yes |
| 3 | * | 1 | 1 | Yes |
| 4 | 0 | 0 | 1 | No |
| | 1 | 1 | 1 | Yes |
| 5 | 0 | 1 | 1 | Yes |
| | 1 | 1 | 1 | Yes |
| 6 | * | 0 | 0 | Yes |
| 7 | * | 0 | 0 | Yes |
| 8 | * | 0 | 0 | Yes |

If our procedure detects a change in period $K$ of a time series, the incrementally built classification model should be modified accordingly. Modification methods could be one of the following [see Last (2002)]: (a) rebuild a new model based on period $K$ only; (b) combine the $DM$ model from periods $1, \ldots, K$, with a new model by using a voting schema (unified, exponential smoothing, any weighted majority, deterministic choice, etc.) in order to improve predictive accuracy in the future; (c) add $j$ more periods in order to rebuild a model based on periods $K, \ldots, K + j$.

In order to illustrate the influence of discarding a detected change, we look at the rules induced from the database by the IFN algorithm before and after the second change has occurred (Period 6).

It can be shown by a simple calculation that if all rules have the same frequency (that is assuming a uniform distribution of all input variables), the expected increase in the error rate as a result of the change in Period 6 is 9% (1 : 11).

## 4. A Real-World Case Study

### 4.1. *Dataset Description*

In this section, we apply our change detection method to a real-world time series dataset. The objectives for the case study are: (1) determine whether or not our method detects changes, which have occurred during some time periods; (2) determine whether or not our method produces "false alarms" while the dataset's characteristics do not change over time significantly.

The dataset has been obtained from a large manufacturing plant in Israel representing daily production orders of products. From now on this dataset will be referred as "*Manufacturing*".

The candidate input attributes are: Catalog number group (CATGRP) — a discrete categorical variable; Market code group (MRKTCODE) — a discrete categorical variable; Customer code group (CUSTOMERGRP) — a discrete categorical variable; Processing duration (DURATION) — a discrete categorical variable which represents the processing times as disjoint intervals of variable size; Time left to operate in order to meet demand (TIME_TO_OPERATE) — a discrete categorical variable which stands for the amount of time between the starting date of the production order and its due date. Each value represents a distinct time interval; Quantity (QUANTITY) — a categorical discrete variable which describes the quantity of items in a production order. Each value represents a distinct quantity interval. The target variable indicates whether the order was delivered on time or not (0 or 1).

The time series database in this case study consists of records of production orders accumulated over a period of several months. The 'Manufacturing' database was extracted from a continuous production sequence. Without further knowledge of the process or any other relevant information about the nature of change of that process, we may assume that no significant changes of the operation characteristics are expected over such a short period of time.

*Presentation and Analysis of Results.* Table 9 and Figure 2 describe the results of applying the IFN algorithm to six consecutive months in the 'Manufacturing' database and using our change detection methodology to detect significant changes that have occurred during these months.

The XP statistics, as described in Table 9 and Figure 2, refer only to the target variable (delivery on time). The magnitude of change in the candidate input variables as evaluated across the monthly intervals is shown in Table 10.

Table 9. Results of the CD hypothesis testing on the 'Manufacturing' database

| Month | CD | | | | | XP |
|---|---|---|---|---|---|---|
| | $e_{M_{K-1},K}$ | $e_{M_{K-1},K-1}$ | $d$ | $H(95\%)$ | $1-p$-value | $1-p$-value |
| 1 | — | — | — | — | — | — |
| 2 | 14.10% | 12.10% | 2.00% | 4.80% | 58.50% | 78.30% |
| 3 | 11.70% | 10.40% | 1.30% | 3.40% | 54.40% | 98.80% |
| 4 | 10.60% | 9.10% | 1.50% | 2.90% | 68.60% | 76.50% |
| 5 | 11.90% | 10.10% | 1.80% | 2.80% | 78.90% | 100.00% |
| 6 | 6.60% | 8.90% | 2.30% | 2.30% | 95.00% | 63.10% |

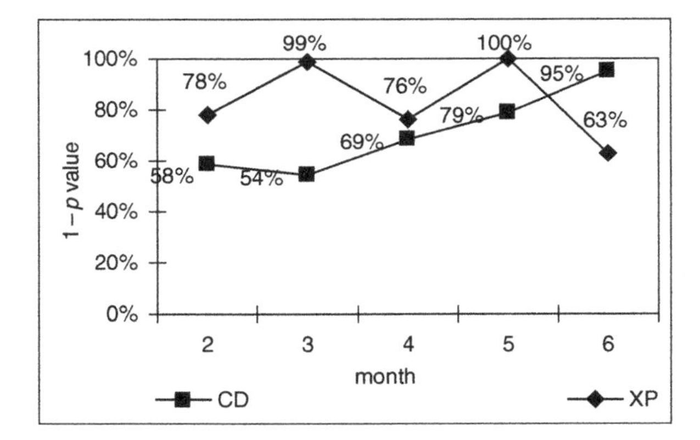


Fig. 2. Summary of implementing the change detection methodology on 'Manufacturing' database ($1 - p$-value).

Table 10. XP confidence level of all independent and dependent variables in 'Manufacturing' database ($1 - p$-value).

| | CAT GRP | MRKT Code | Duration | Time to operate | Quantity | Customer GRP |
|---|---|---|---|---|---|---|
| Domain | 18 | 19 | 19 | 19 | 15 | 18 |
| Month 2 | 100% | 100% | 100% | 100% | 100% | 100% |
| Month 3 | 100% | 100% | 100% | 100% | 100% | 100% |
| Month 4 | 100% | 99.8% | 100% | 100% | 100% | 100% |
| Month 5 | 100% | 99.9% | 100% | 100% | 100% | 100% |
| Month 6 | 100% | 100% | 100% | 100% | 100% | 100% |

According to the change detection methodology, during all six consecutive months there was no significant change in the rules describing the relationships between the candidate and the target variables (which is our main interest). Nevertheless, it is easy to notice that major changes have been revealed by the XP statistic in distributions of most target and candidate input variables. One can expect that variables with a large number of values need greater data sets in order to reduce the variation of their distribution across periods. However, this phenomenon has not affected the CD statistic.

An interesting phenomenon is the increasing rate of the CD confidence level from month 2 to month 6. In order to further investigate whether a change in frequency distribution has still occurred during the six consecutive months without resulting in a significant CD confidence level, we have validated the sixth month on the fifth and the first months. Table 11 and Figure 3 describe the outcomes of the change detection methodology.

Implementing the change detection methodology by validating the sixth month on the fifth and the first month did not produce contradicting results. That is, the CD confidence level of both months ranges only within $\pm 8\%$ from the original CD estimation based on the all five previous months. Furthermore, although XP produced extremely high confidence levels indicating a drastic change in the distribution of all candidate and target variables, the data mining model was not affected, and it kept producing similar validation error rates (which were statistically evaluated by CD).

The following statements summarize the case study's detailed results:

- Our expectation that in the 'Manufacturing' database, there are no significant changes in the relationship between the candidate input variables and the target variable over time, is validated by the change detection

Table 11. Outcomes of XP by validating the sixth month on the fifth and the first month in 'Manufacturing' database ($p$-value).

| | | CAT GRP | MRKT Code | Duration | Time to Operate | Quantity | Customer GRP | Target |
|---|---|---|---|---|---|---|---|---|
| Metric XP | domain | 18 | 19 | 19 | 19 | 15 | 18 | 2 |
| ($1 - p$-value) | month 5 validated by month 6 | 100% | 100% | 100% | 100% | 100% | 100% | 98.4% |
| | month 1 validated by month 6 | 100% | 100% | 100% | 100% | 100% | 100% | 100% |
| | months 1 to 5 validated by month 6 | 100% | 100% | 100% | 100% | 100% | 100% | 63.1% |

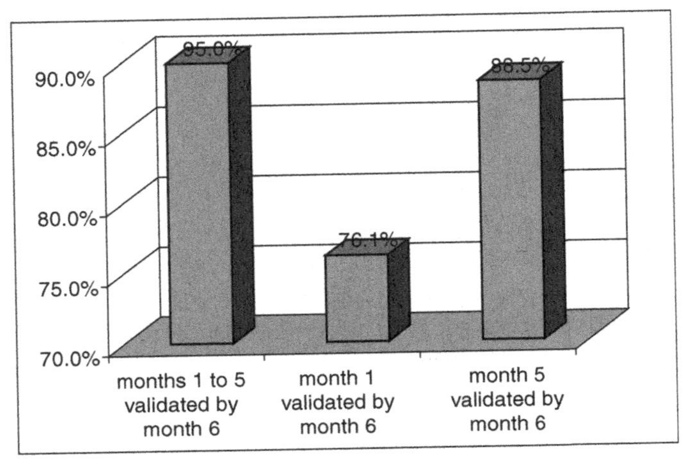


Fig. 3. CD confidence level ($1 - p$-value) outcomes of validating the sixth month on the fifth and the first month in 'Manufacturing' database.

procedure. No results of the changed detection metric (CD) exceeding the 95% confidence level were produced in any period. This means that no "false alarms" were issued by the procedure.

- Statistically significant changes in the distributions of the candidate input (independent) variables and the target (dependent) variable across monthly intervals have not generated a significant change in the rules, which are induced from the database.
- The CD metric implemented by our method can also be used to determine whether an incrementally built model is stable. If we are applying a stable data mining algorithm, like the Info-Fuzzy Network, to an accumulated amount of data, it should produce increasing confidence levels of the CD metric over the initial periods of the time series, as more data supports the induced classification model.

Thus the results obtained from a real-world time series database confirm the conclusions of the experiments on artificial datasets with respect to reliability of the proposed change detection methodology.

## 5. Conclusions and Future Work

As mentioned above, most methods of batch learning are based on the assumption that the training data involved in building and verifying the model is the best estimator of what will happen in the future. An important factor that must not be set aside is the time factor. As more data is accumulated in a time series database, incrementally over time, one must examine whether the data in a new period agrees with the data in previous periods and take the relevant decisions about learning in the future. This work presents a novel change detection method for detecting significant changes in classification models induced from continuously accumulated data streams by batch learning methods.

The following statements can summarize the major contributions of this work to the area of data mining and knowledge discovery in databases:

(i) This work defines three main causes for a statistically significant change in a data-mining model:

- A change in the probability distribution of one or more of candidate input variables $A$.
- A change in the distribution of the target variable $T$.
- A change in the "patterns" (rules), which define the relationship of the candidate input to the target variable. That is, a change in the model $M$.

This work has shown that although there are three main causes for significant changes in the data-mining models, it is common that these main causes co-exist in the same data stream, deriving eight possible combinations for a significant change in a classification model induced from time series data. Moreover, these causes affect each other in a manner and magnitude that depend on the database being mined and the algorithm in use.

(ii) The change can be detected by the change detection procedure using a three-stage validation technique. This technique is designed to detect all possible significant changes.

(iii) The change detection method relies on the implementation of two statistical tests:

(a) Change Detection hypothesis testing ($CD$) of every period $K$, based on the definition of a significant change $C$ in classification "rules", with respect to the previous $K - 1$ periods.

(b) Pearson's estimator ($XP$) for testing matching proportions of variables to detect a significant change in the probability distribution of candidate input and target attributes.

(iv) The effect of a change is relevant and can be mainly detected in the period of change. If not detected, the influence of a change in a database can be absorbed in successive periods.
(v) All procedures, hypothesis tests and definitions, were validated on artificially generated databases and on a real-world database.
(vi) The change detection procedure has low computational costs. Its complexity is $O(n_K)$, where $n_K$ is the total number of validation records in $K$ periods, since it requires only testing whether the new data agrees with the model induced from previously aggregated data.
(vii) The CD metric can also be used to determine whether an incrementally built model is stable. In our real-world experiment, stability of info-fuzzy network has been confirmed by the increasing confidence levels over initial periods of the data stream.

The Change Detection Procedure, with the use of the statistical estimators, can detect significant changes in classification models of data mining. These changes can be detected independently of the data mining algorithm used (e.g., C4.5 and ID3 by Quinlan, KS2, ITI and DMTI by Utgoff, IFN by Last and Maimon, IDTM by Kohavi, Shen's CDL4, etc. ), or the induced classification model (rules, decision trees, networks, etc.). As change detection is quite a new application area in the field of data mining, many future issues could be developed, including the following:

(i) Implementing meta-learning techniques according to the cause(s) and magnitude(s) of a change(s) detected in period $K$ for combining several models, such as: exponential smoothing, voting weights based on the *CD* confidence level, ignoring old or problematic periods, etc.
(ii) Increasing the granularity of significant changes. There should be several sub-types of changes with various magnitudes of the model structure and parameters that could be identified by the change detection procedure, in order to give the user extra information about the mined time series.
(iii) Integrating the change detection methodology in an existing data mining algorithm. As indicated above, the change detection procedure's complexity is only $O(n)$. One can implement this procedure in an existing incremental (online) learning algorithm, which will continue efficiently rebuilding an existing model if the procedure does not indicate a significant change in the newly obtained data. This option is also applicable for meta-learning and multi-model methods.

(iv) Using the CD statistical hypothesis testing for continuous monitoring of specific attributes.
(v) Using the CD statistical estimator or its variations as a metric for measuring the stability of an incrementally built data mining model.
(vi) Further analysis of the relationship between the two error rates $\alpha$ (false positive) and $\beta$ (false negative) and its impact on the performance of the change detection procedure.
(vii) Further validation of the Change Detection Procedure on other $DM$ models, algorithms, and data types.

## References


1. Ali, K. and Pazzani, M. (1995). Learning Multiple Descriptions to Improve Classification Accuracy. *International Journal on Artificial Intelligence Tools*, **4**, 1–2.
2. Case, J., Jain, S., Lange, S. and Zeugmann, T. (1999). Incremental Concept Learning for Bounded Data Mining *Information & Computation*, **152**(1), pp. 74–110.
3. Chan, P.K. and Stolfo, S.J. (1995). A Comparative Evaluation of Voting and Meta-learning on Partitioned Data. *Proceedings of the Twelfth International Conference on Machine Learning*, pp. 90–98.
4. Chan, P.K. and Stolfo, S. (1996). Sharing Learned Models among Remote Database Partitions by Local Meta-learning. *Proc. Second Intl. Conf. on Knowledge Discovery and Data Mining*, pp. 2–7.
5. Cheung, D., Han, J., Ng, V., and Wong, C.Y. (1996). Maintenance of Discovered Association Rules in Large Databases: An Incremental Updating Technique. *Proceedings of 1996 International Conference on Data Engineering (ICDE'96)*, New Orleans, Louisiana, USA, Feb. 1996.
6. Domingos, P. (1997). Bayesian Model Averaging in Rule Induction. Preliminary Papers of the *Sixth International Workshop on Artificial Intelligence and Statistics*, Ft. Lauderdale, FL: Society for Artificial Intelligence and Statistics, pp. 157–164.
7. Domingos, P. (1997). Knowledge Acquisition from Examples Via Multiple Models. *Proceedings of the Fourteenth International Conference on Machine Learning*, Nashville, TN: Morgan Kaufmann, pp. 98–106.
8. Domingos, P. (1996). Using Partitioning to Speed Up Specific-to-General Rule Induction. *Proceedings of the AAAI-96 Workshop on Integrating Multiple Learned Models,* AAAI Press, pp. 29–34.
9. Fawcett, T. and Provost, F. (1999). Activity Monitoring: Noticing Interesting Changes in Behavior. *Proceedings of the Fifth International Conference on Knowledge Discovery and Data Mining (KDD-99).*
10. Fayyad, U., Piatetsky-Shapiro, G., and Smyth, P. (1996). Knowledge Discovery and Data Mining: Towards a Unifying Framework. *Proceedings of the Second ACM SIGKDD International Conference on Knowledge Discovery and Data Mining (KDD-96).*

11. Freund, Y. and Mansour, Y. (1997). Learning under persistent drift. *Proceedings of EuroColt*, pp. 109–118.
12. Gerevini, A., Perini, A., and Ricci, F. (1996). Incremental Algorithms for Managing Temporal Constraints. *IEEE International Conference on Tools with Artificial Intelligence (ICTAI'96)*.
13. Harries, M. and Horn, K. (1996). Learning stable concepts in domains with hidden changes in context. *Proceedings of the ICML-96 Workshop on Learning in Context-Sensitive Domains*, Bari, Italy, July 3, pp. 21–31.
14. Helmbold, D.P. and Long, P.M. (1994). Traching Drifting Concepts By Minimizing Disagreements. *Machine Learning*, **14**(1), 27–46.
15. Hsu, C.-N. and Knoblock, C.A. (1998). Discovering Robust Knowledge from Databases that Change. *Journal of Data Mining and Knowledge Discovery*, **2**(1), 1–28.
16. Hines, W.H. and Montgomery, D.C. (1990). *Probability and Statistics in Engineering and Management Science.* Third Edition, Wiley.
17. Jones, R.H., Crowell, D.H., and Kapuniai, L.E. (1970). Change detection model for serially correlated multivariate data. *Biometrica*, **26**, 269–280.
18. Hulten, G., Spencer, L., and Domingos, P. (2001). Mining Time-Changing Data Streams. *Proceedings of the Seventh ACM SIGKDD International Conference on Knowledge Discovery and Data Mining (KDD-2001)*.
19. Kelly, M.G., Hand, D.J., and Adams, N.M. (1999). The Impact of Changing Populations on Classifier Performance. *Proceedings of the Fifth ACM SIGKDD International Conference on Knowledge Discovery and Data Mining (KDD-99).*
20. Kohavi, R. (1995). The Power of Decision Tables. *Proceedings of European Conference on Machine Learning (ECML).*
21. Lane, T. and Brodley, C.E. (1998). Approaches to Online and Concept Drift for User Identification in Computer Security. *Proceedings of the Fourth International Conference on Knowledge Discovery and Data Mining*, New York, NY, pp. 259–263.
22. Last, M., Maimon, O., and Minkov, E. (2002). Improving Stability of Decision Trees. *International Journal of Pattern Recognition and Artificial Intelligence*, **16**(2), 145–159.
23. Last, M. (2002). Online Classification of Non-stationary Data Streams. *Intelligent Data Analysis*, **6**(2), 129–147.
24. Liu, H. and Setiono, R. (1998). Incremental Feature Selection. *Journal of Applied Intelligence*, **9**(3), 217–230.
25. Maimon, O. and Last, M. (2000). *Knowledge Discovery and Data Mining, the Info-Fuzzy Network (IFN) Methodology*, Kluwer.
26. Martinez, T. (1990). Consistency and Generalization in Incrementally Trained Connectionist Networks. *Proceeding of the International Symposium on Circuits and Systems*, pp. 706–709.
27. Mangasarian, O.L. and Solodov, M.V. (1994). Backpropagation Convergence via Deterministic Nonmonotone Perturbed Mininization. *Advances in Neural Information Processing Systems*, **6**, 383–390.

28. Minium, E.W., Clarke, R.B., and Coladarci, T. (1999). *Elements of Statistical Reasoning*, Wiley, New York.
29. Mitchell, T.M. (1997). *Machine Learning*, McGraw Hill.
30. Montgomery, D.C. and Runger, G.C. (1999). *Applied Statistics and Probability for Engineers*, Second Edition, Wiley.
31. Nouira, R. and Fouet, J.-M. (1996). A Knowledge Based Tool for the Incremental construction, Validation and Refinement of Large Knowledge Bases. *Preliminary Proceedings of Workshop on validation, verification and refinement of BKS (ECAI96)*.
32. Ohsie, D., Dean, H.M., Stolfo, S.J., and Silva, S.D. (1995). Performance of Incremental Update in Database Rule Processing.
33. Shen, W.-M. (1997). An Active and Semi-Incremental Algorithm for Learning Decision Lists. *Technical Report*, USC-ISI-97, Information Sciences Institute, University of Southern California, 1997. Available at: http://www.isi.edu/~shen/active-cdl4.ps.
34. Shen, W.-M. (1997). Bayesian Probability Theory — A General Method for Machine Learning. From *MCC-Carnot-101-93. Microelectronics and Computer Technology Corporation*, Austin, TX. Available at: http://www.isi.edu/~shen/Bayes-ML.ps.
35. Utgoff, P.E. and Clouse, J.A. (1996). A Kolmogorov-Smirnoff metric for decision tree induction. *Technical Report 96-3*, Department of computer science, University of Massachusetts. Available at: ftp://ftp.cs.umass.edu/pub/techrept/techreport/1996/UM-CS-1996-003.ps
36. Utgoff, P.E. (1994). An improved algorithm for incremental induction of decision trees. *Machine Learning: Proceedings of the Eleventh International Conference*, pp. 318–325.
37. Utgoff, P.E. (1995). Decision tree induction based on efficient tree restructuring. *Technical Report* 95-18, Department of computer science, University of Massachusetts. Available at: ftp://ftp.cs.umass.edu/pub/techrept/techreport/1995/UM-CS-1995-018.ps
38. Widmer, G. and Kubat, M. (1996). Learning in the Presence of Concept Drift and Hidden Contexts. *Machine Learning*, **23**(1), 69–101.
39. Yao, Y. (1988). Estimating the number of change points via Schwartz' criterion. *Statistics and probability letters*, pp. 181–189.
40. Zhang, B.-T. (1994). An Incremental Learning Algorithm that Optimizes Network Size and Sample Size in One Trial. *Proceedings of International Conference on Neural Networks (ICNN-94)*, IEEE, pp. 215–220.

# CHAPTER 6

# CLASSIFICATION AND DETECTION OF ABNORMAL EVENTS IN TIME SERIES OF GRAPHS

H. Bunke
*Institute of Computer Science and Applied Mathematics*
*University of Bern, CH-3012 Bern, Switzerland*
E-mail: bunke@iam.unibe.ch

M. Kraetzl
*ISR Division, Defense Science and Technology Organization*
*Edinburgh SA 5111, Australia*
E-mail: mkraetz@nsa.gov

Graphs are widely used in science and engineering. In this chapter, the problem of detecting abnormal events in times series of graphs is investigated. A number of graph similarity measures are introduced. These measures are useful to quantitatively characterize the degree of change between two graphs in a time series. Based on any of the introduced graph similarity measures, an abnormal change is detected if the similarity between two consecutive graphs in a time series falls below a given threshold. The approach proposed in this chapter is not geared towards any particular application. However, to demonstrate its feasibility, its application to abnormal event detection in the context of computer networks monitoring is studied.

*Keywords*: Graph; time series of graphs; graph similarity; abnormal event detection; computer network monitoring.

## 1. Introduction

Graphs are a powerful and flexible data structure useful for the representation of objects and concepts in many disciplines of science and engineering. In a graph representation, the nodes typically model objects, object parts, or object properties, while the edges describe relations between the nodes, for example, temporal, spatial, or conceptual dependencies between the objects that are modelled through the nodes. Examples of applications where graph representations have been successfully used include chemical structure analysis [1], molecular biology [2], software engineering [3], and database retrieval [4]. In artificial intelligence, graphs have been successfully applied to case-based reasoning [5], machine learning [6], planning [7], knowledge representation [8], and data mining [9]. Also in computer vision and pattern recognition graphs are widely used. Particular applications include character recognition [10], graphics recognition [11], shape analysis [12], and object classification [13].

In many artificial intelligence applications object representations in terms of feature vectors or lists of attribute-value pairs are used [14]. On the one hand, graphs have a much higher representational power than feature vectors and attribute-value lists as they are able to model not only unary, but higher order relations and dependencies between various entities. On the other hand, many operations on graphs are computationally much more costly than equivalent operations on feature vectors and attribute-value lists. For example, computing the dissimilarity, or distance, of two objects that are represented through feature vectors is an operation that is linear in the number of features, but computing the edit distance of two graphs in exponential in the number of their nodes [15].

In this chapter we focus on a special class of graphs that is characterized by a low computational complexity with regard to a number of operations. Hence this class of graphs offers the high representational power of graphs together with the low computational complexity typically found with feature vector representations. The computational efficiency of the graph operations results from constraining the graphs to have unique node labels. In particular we are concerned with time series of graphs. In such series, each individual graph represents the state of an object, or a system, at a particular point of time. The task under consideration is the detection of abnormal events, i.e. the detection of abnormal change of the state of the system when going from one point in time to the next.

Data mining is generally concerned with the detection and extraction of meaningful patterns and rules from large amounts of data [16,17]. Classification is considered to be one of the main subfields of data mining [18]. The goal of classification is to assign an unknown object to one out of a given number of classes, or categories. Obviously, the task considered in the present paper is a particular instance of classification, where the transitions between the individual elements of a given sequence of graphs are to be classified as normal or abnormal. From the general point of view, such a classification may serve as the first step of a more complex data mining and knowledge extraction process that is able to infer hitherto hidden relationships between normal and abnormal events in time series.

The procedures introduced in this chapter do not make any particular assumptions about the underlying problem domain. In other words, the given sequence of graphs may represent any time series of objects or systems. However, to demonstrate the feasibility of the proposed approach, a special problem in the area of computer network monitoring is considered [19]. In this application each graph in a sequence represents a computer network at a certain point of time, for example, at a certain time each day. Changes of the network are captured using a graph distance measure. Hence a sequence of graph similarity values is derived from a given time series of graphs. Each similarity value is an indicator of the degree of change that occurred to the network between two consecutive points of time. In this sequence of similarity values abnormal change can be detected by means of thresholding and similar techniques. That is, it is assumed that an abnormal change has occurred if the similarity between two consecutive graphs is below a given threshold.

There are other applications where complex systems, which change their behaviour or properties over time, are modelled through graphs. Examples of such systems include electrical power grids [20], regulatory networks controlling the mammalian cell cycle [21], and co-authorship and citation networks in science [22]. The formal tools introduced in this chapter can also be used in these applications to detect abnormal change in the system's behaviour.

The remainder of this chapter is organized as follows. In Section 2 our basic terminology is introduced. Graph similarity measures based on graph spectra are presented in Section 3. Another approach to measuring graph similarity, using graph edit distance, is discussed in Section 4. The median of a set, or sequence, of graphs is introduced in Section 5, and its use for the classification of events in a sequence of graphs is discussed in Section 6. Then in Section 7 the application of the similarity measures introduced in the previous sections to the detection of abnormal change in communication networks is discussed. Finally, some conclusions are drawn in Section 8.

## 2. Preliminaries

A *graph* $G = (V, E)$ consists of a finite set of vertices $V$ and finite set of edges $E$ which are pairs of vertices; a pair of vertices denotes the endpoints of an edge. Two vertices $u, v \in V$ are said to be *adjacent* if they are endpoints of the same edge [23]. Edges in $G$ can be *directed.* In this case edge $(u, v) \in E$ originates at node $u \in V$ and terminates at node $v \in V$.

Objects such as vertices or edges (or their combinations) associated with a graph are referred to as *elements* of the graph. A *weight* is a function whose domain is a set of graph elements in $G$. The domain can be restricted to that of edge or vertex elements only, where the function is referred to as *edge-weight* or *vertex-weight* respectively. A weight whose domain is all vertices and edges is called *total. Values* of weight assigned to elements in $G$ may be numerical (e.g. the amount of traffic in a computer network as values of edge-weight), or symbolic (e.g. node identifiers as values of vertex-weight). The set of possible values in the range of the weight function are called *attributes.* A unique labeling is a one-to-one weight, for a vertex-weight this serves to uniquely identify vertices of a graph. A graph with a unique labeling is called a *labeled* graph.

The graph $G = (V, E, w_V, w_E)$ is vertex-labeled with weight $w_V\colon V \to L_V$ assigning vertex identifier attributes $L_V$ to individual vertices, where the value of an assigned vertex-weight is $w_V(u)$ for vertex $u$. Furthermore, $w_V(u) \neq w_V(v), \forall u, v \in V, uv$ since the graphs in our application possess unique vertex-labelings. Edges are also weighted with weight $w_E\colon E \to R^+$. The notation $w_E^G$ is used to indicate that edge weight $w_E$ belongs to graph $G$. The number of vertices in $G = (V, E)$ is denoted by $|V|$, and likewise the number of edges is denoted by $|E|$.

## 3. Analysis of Graph Spectra

One of the known approaches to the measurement of change in a time series of graphs is through the analysis of graph spectra. Algebraic aspects of spectral graph theory are useful in the analysis of graphs [24–27]. There are several ways of associating matrix spectra (sets of all eigenvalues) with a given weighted graph $G$. The most obvious way is to investigate the structure of a finite digraph $G$ by analyzing the spectrum of its adjacency matrix $\mathbf{A}_G$. Note that $\mathbf{A}_G$ is not a symmetric matrix for directed graphs because weight may not be a symmetric function.

For a given ordering of the set of $n$ vertices $V$ in a graph $G = (V, E)$, one can investigate the *spectrum* $\sigma(G) = \{\lambda_1, \lambda_2, \ldots, \lambda_n\}$, where $\lambda_i$ are the eigenvalues of the weighted adjacency matrix $\mathbf{A}_G$. Obviously, $\sigma(G)$ does not depend on the ordering of $V$. Also, the matrix spectrum does not change under any nonsingular transformation. Since the adjacency matrices have nonnegative entries, the class of all isospectral graphs (graphs having identical spectra) must be relatively small. Because of that fact, and because isomorphic graphs are isospectral, one might expect that classes of isospectral and isomorphic graphs coincide. However, it can be shown that the class of isospectral graphs is larger [24].

It is easy to verify that by (arbitrary) assignment of $x_i$'s in an eigenvector $\mathbf{x} = [x_1, x_2, \ldots, x_n]$ of $\mathbf{A}_G$ to $n$ various vertices in $V$, the components of $x$ can be interpreted as positive vertex-weight values (attributes) of the corresponding vertices in the digraph $G$. This property is extremely useful in finding connectivity components of $G$, and in various clustering, coarsening, or graph condensing implementations [28].

A different approach to graph spectra for undirected graphs [25,29,30] investigates the eigenvalues of the *Laplace matrix* $\mathbf{L}_G = \mathbf{D}_G - \mathbf{A}_G$, where the *degree matrix* $\mathbf{D}_G$ of graph $G$ is defined as $\mathbf{D}_G = \mathrm{diag}\{\sum_{v \in V} w_E^G(u,v) | u \in V_G\}$. Note that in the unweighted case, diagonal elements of $\mathbf{D}_G$ are simply the vertex degrees of indexed vertices $V$. It follows that the Laplacian spectrum is always nonnegative, and that the number of zero eigenvalues of $\mathbf{L}_G$ equals the number of connectivity components of $G$. In the case of digraphs, the Laplacian is defined as $\mathbf{L}_G = \mathbf{D}_G - (\mathbf{A}_G + \mathbf{A}_G^T)$ (in order to ensure that $\sigma(\mathbf{L}_G) \subseteq \{0\} \cup \mathbf{R}^+$). The second smallest eigenvalue (called the *algebraic connectivity* of $G$) provides graph connectivity information and is always smaller or equal to the vertex connectivity of $G$ [30]. Laplacian spectra of graphs have many applications in graph partitions, isoperimetric problems, semidefinite programming, random walks on graphs, and infinite graphs [25,29].

The relationship between graph spectra and graph distance measures can be established using an eigenvalue interpretation [27]. Consider the weighted graph matching problem for two graphs $G = (V_G, E_G)$ and $H = (V_H, E_H)$, where $|V_G| = |V_H| = n$, with positive edge-weights $w_E^G$ and $w_E^H$, which is defined as finding a one-to-one function $\Phi\colon V_G \to V_H$ such that the *graph distance function*:

$$\mathrm{dist}(\Phi) = \sum_{u \in V_G, v \in V_H} \left[ w_E^G(u,v) - w_E^H(\Phi(u), \Phi(v)) \right]^2 \tag{3.1}$$

is minimal. This is equivalent to finding a permutation matrix $\mathbf{P}$ that minimises $J(\mathbf{P}) = ||\mathbf{P}\mathbf{A}_G\mathbf{P}^T - \mathbf{A}_H||^2$, where $\mathbf{A}_G$ and $\mathbf{A}_H$ are the (weighted) adjacency matrices of the two graphs, and $||\cdot||$ is the ordinary Euclidean matrix norm. Notice that this graph matching problem is computationally expensive because it is a combinatorial optimisation problem.

In the case of two graphs $G$ and $H$ with the same vertex set $V$, and (different) edge-weights $w_E^G$ and $w_E^H$, the *(Umeyama) graph distance* is expressed as:

$$\text{dist}(G,H) = \sum_{u,v\in V} \left[w_E^G(u,v) - w_E^H(u,v)\right]^2. \tag{3.2}$$

In spite of the computational difficulty of finding a permutation matrix $\mathbf{P}$ which minimises $J(\mathbf{P})$, in the restricted case of orthogonal (permutation) matrices, one can use the spectra (eigendecompositions) of the adjacency matrices $\mathbf{A}_G$ and $\mathbf{A}_H$. In fact, for Hermitian matrices $\mathbf{A},\mathbf{B} \in M_{nn}(\mathbf{R})$ with corresponding spectra $\sigma(\mathbf{A}) = \{\alpha_1 > \alpha_2 > \cdots > \alpha_n\}$ and $\sigma(\mathbf{B}) = \{\beta_1 > \beta_2 > \cdots > \beta_n\}$, and with eigendecompositions $\mathbf{A} = \mathbf{U}_A\mathbf{D}_A\mathbf{U}_A^T$ and $\mathbf{B} = \mathbf{U}_B\mathbf{D}_B\mathbf{U}_B^T$($\mathbf{U}_{(\cdot)}$ are unitary, and $\mathbf{D}_{(\cdot)}$ are diagonal matrices):

$$\sum_{i=1}^{n}(a_i - b_i)^2 = \min_{\mathrm{Q}} \parallel \mathbf{QAQ}^T - \mathbf{B} \parallel^2, \tag{3.3}$$

where the minimum is taken over all unitary matrices $\mathbf{Q}$. Equation (3.3) justifies the use of the adjacency matrix spectra; given two weighted graphs $G$ and $H$ with respective spectra

$$\sigma(\mathbf{A}_G) = \{\lambda_1, \lambda_2, \ldots, \lambda_{n_1}\} \quad \text{and} \quad \sigma(\mathbf{A}_H) = \{\mu_1, \mu_2, \ldots, \mu_{n_2}\},$$

the $k$ largest positive eigenvalues are incorporated into the graph distance measure [26] as:

$$\text{dist}(G,H) = \sqrt{\frac{\sum_{i=1}^{k}(\lambda_i - \mu_i)^2}{\min\left\{\sum_{i=1}^{k}\lambda_i^2, \sum_{j=1}^{k}\mu_j^2\right\}}}, \tag{3.4}$$

where $k$ is an arbitrary summation limit, but empirical studies in pattern recognition and image analysis show that $k \approx 20$ is a good choice. Notice that similar approaches to distance measures can be applied to the case of Laplacian spectra.

## 4. Graph Edit Distance

In this section we introduce a graph distance measure, termed graph edit distance **ged**, that is based on graph edit operations. This distance measure evaluates the sequence of edit operations required to modify an input graph such that it becomes isomorphic to some reference graph. This can include the possible insertion and deletion of edges and vertices, in addition to weight value substitutions [31]. Generally, **ged** algorithms assign costs to each of the edit operations and use efficient tree search techniques to identify the sequence of edit operations resulting in the lowest total edit cost [15,32]. The resultant lowest total edit cost is a measure of the distance between the two graphs.

In general graph matching problems with unlabeled graphs, a unique sequence of edit operations does not exist due to the occurrence of multiple possible vertex mappings. The ged algorithms are required to search for the edit sequence that results in a minimum edit cost. However, for the class of graphs introduced in Section 2, vertex-weight value substitution is not a valid edit operation because vertex-weight values are unique. As a result, the combinatorial search reduces to the simple identification of elements (vertices and edges) inserted or deleted from one graph $G$ to produce the other graph $H$. The implementation requires linear time in size of the problem.

If the cost associated with the insertion or deletion of individual elements is one, and edge-weight value substitution is not considered (i.e., we consider the topology only), the edit sequence cost becomes the difference between the total number of elements in both graphs, and all graph elements in common.

Using the above cost function, let $G = \left(V_G, E_G, w_V^G, w_V^G\right)$ and $H = \left(V_H, E_H, w_V^H w_V^H\right)$ be two graphs the similarity of which is to be evaluated. The *graph edit distance* $d_1(G, H)$ describing topological change that takes place when going from graph $G$ to $H$ then becomes:

$$d_1(G, H) = |V_G| + |V_H| - 2|V_G \cap V_H| + |E_G| + |E_H| - 2|E_G \cap E_H|. \quad (4.1)$$

Clearly the edit distance, as a measure of topology change, increases with increasing degree of change. Edit distance $d_1(G, H)$ is bounded below by $d_1(G, H) = 0$ when $H$ and $G$ are isomorphic (i.e., there is no change), and above by $d_1(G, H) = |V_G| + |V_H| + |E_G| + |E_H|$ when $G \cup H = \emptyset$, the case where the graphs are completely different.

In the second graph edit distance measure studied in this paper, we consider not only change in graph topology, but also in edge weight. For this purpose, we assign a cost $c$ to each vertex deletion and insertion, where $c > 0$ is a constant. The cost of changing weight $w_E^G(e)$ on edge $e \in E_G$ into $w_E^H(e)$ on $e \in E_H$ is defined as $\left|w_E^G(e) - w_E^H(e)\right|$. To simplify our notation we let our graphs be completely connected (i.e., there is an edge $e \in E_H$ between any two vertices in $G$ and an edge $e \in E_H$ between any two vertices in $H$) and assign a weight equal to zero to edge $e \in E_G$ ($e \in E_H$) if this edge does not exist in $G(H)$. Hence substitution of edge weights, edge deletions, and edge insertions can be treated uniformly. Note that the deletion of an edge $e \in E_G$ with weight $w_E^G(e)$ has a cost equal to $w_E^G(e)$. Similarly the insertion of an edge $e \in E_H$ has the weight of that edge, $w_E^H(e)$, assigned as its cost. Consequently, the graph edit distance under this cost function becomes

$$d_2(G,H) = c[|V_G| + |V_H| - 2|V_G \cap V_H|] + \sum_{e \in E_G \cap E_H} |w_E^G(e) - w_E^H(e)| + \sum_{e \in E_G \setminus (E_G \cap E_H)} w_E^G(e) + \sum_{e \in E_H \setminus (E_G \cap E_H)} w_E^H(e). \quad (4.2)$$

The constant $c$ is a parameter that allows us to weight the importance of a vertex deletion or insertion relatively to the weight changes on the edges.

## 5. Median Graphs

Intuitively speaking, the *median* of a sequence of graphs $S = (G_1, \ldots, G_n)$ is a single graph that represents the given $G_i$'s in the best possible manner. Using any of the available graph distance measures, for example, $d_1(G,H)$ or $d_2(G,H)$ introduced in Section 4, the median of a sequence of graphs $S$ is defined as a graph that minimises the sum of the edit distances to all members of sequence $S$. Formally, let $U$ be the family of all graphs that can be constructed using labels from $L_V$ for vertices and real numbers for edges. Then $\bar{G}$ is a *median graph* of the sequence $S = (G_1, \ldots, G_n)$ if:

$$\bar{G} = \arg\min_{G \in U} \sum_{i=1}^{n} d(G, G_i). \quad (5.1)$$

If we constrain $\bar{G}$ to be a member of $S$ then the graph that satisfies Eq. (5.1) is called the *set median* of $S$. In a general context where node labels are not unique a set median is usually easier to compute than a median graph. However, in the context of the present paper where we have uniquely labelled graphs, we will exclusively focus on median graph computation and its application to abnormal change detection. It is to be noted that median graphs need not be unique.

The term "median graph" is used because the median $\bar{x}$ of an ordered sequence of real numbers $(x_1, \ldots, x_n)$, which is defined as

$$\text{if } n \text{ is even, then } \bar{x} = \text{median}(x_1, \ldots, x_n) = x_{n/2},$$
$$\text{else } \bar{x} = \text{median}(x_1, \ldots x_n) = x_{\lceil n/2 \rceil}$$

has a similar property as the median of a sequence of graphs: it minimises the sum of distances to all elements of the given sequence i.e., it minimises the expression $\sum_{i=1}^{n} |\bar{x} - x_i|$.

Next we describe a procedure for the computation of the median $G$ of a sequence of graphs $(G_1, \ldots, G_n)$ using the topological graph distance measure $d_1$ defined in Eq. (4.1). Let $G = (V, E)$ with $V = \bigcup_{i=1}^{n} V_i$ and $E = \bigcup_{i=1}^{n} E_i$. Furthermore, let $C(u)$ denote the total number of occurrences of vertex $u$ in $V_1, \ldots, V_n$. Formally, $C(u)$ is defined by the following procedure:

```
C(u) = 0;
for i = 1 to n do
if u ∈ V_i then C(u) = C(u) + 1
```

An analogous procedure can be defined for the edges $(u, v) \in E$:

```
C(u,v) = 0;
for i = 1 to n do
if (u,v) ∈ E_i then C(u,v) = C(u,v) + 1
```

Next we define graph $\bar{G} = (\bar{V}, \bar{E})$ where:

$$\bar{V} = \{u | u \in V \wedge C(u) > n/2\},$$
$$\bar{E} = \{(u, v) | (u, v) \in E \wedge C(u, v) > n/2\}.$$

Then it can be proven that $\bar{G}$ is a median of sequence $(G_1, \ldots, G_n)$ [33]. Consequently, a practical procedure for computing the median of sequence $S$ of graphs is as follows. We consider the union of all vertices and all edges in $S$. For each vertex $u$ a counter, $C(u)$, and for each edge $(u, v)$ a counter, $C(u, v)$, is defined. For each instance of vertex $u$ (and edge $(u, v)$) in sequence $S$, $C(u)$ (and $C(u, v)$) is incremented by one. Finally, vertex $u$ (edge $(u, v)$) is included in $\bar{G}$ if $C(u) > n/2$ ($C(u, v) > n/2$). Obviously, this procedure is linear in the number of vertices and edges, and in the number of graphs in $S$. In contrast to the general method discussed in [34], which is computationally very costly, it can be expected that the procedure introduced in this paper can handle long sequences of large graphs.

In [34] it was shown that the median of a set of graphs is not necessarily unique. It is easy to see that the median graph computed by the procedure introduced in this section is always unique. However, note that the condition $C(u) > n/2$ and $C(u, v) > n/2$ can be relaxed by inclusion of a vertex, or an edge, in $G$ if $C(u) = n/2$, or $C(u,v) = n/2$, respectively. Obviously, any graph resulting from this relaxed procedure is also a median of sequence $S$. Consequently, for the graph distance measure $d_1$ used in this section, several medians for a given sequence of graphs may exist. They result from either inclusion or non-inclusion of a vertex $u$ (or an edge $(u,v)$) with $C(u) = n/2$ (or $C(u,v) = n/2$) in $G$.

An example of the situation where the median graph of a sequence is not unique is shown in Figure 1. Here we have the choice for both vertex 0 and 3 and their incident edges of either including or not in the median graph. This leads to a total of nine possible median graphs depicted in Figure 2.

Another property worth mentioning is the convergence property of the median under distance $d_1$. That is, for any given sequence $S = (G_1, \ldots, G_n)$ of graphs,

$$\text{median}(G_1, \ldots, G_n) = \text{median}(G_1, \ldots, G_n, \text{median}(G_1, \ldots, G_n)).$$

Clearly, each vertex $u$ and each edge $(u,v)$ occurring in sequence $S$ is included in the median if and only if it occurs in $k > n/2$ graphs. Hence, adding its median number of occurrences to set $S$ will result in a median that is identical to the median of $S$. For example, $\bar{G}_1$ shown in Figure 2 is a median of $(G_1, G_2)$ in Figure 1, but $\bar{G}_1$ is also a median of the sequence $(\bar{G}_1, G_1, G_2)$.

In the remainder of this section we derive a procedure for median graph computation using the generalized graph distance measure $d_2$ defined in Eq. (4.2). We start again from $S = (G_1, \ldots, G_n)$, $G = (V, E)$ with $V = \bigcup_{i=1}^n V_i$ and $E = \bigcup_{i=1}^n E_i$, and let $C(u)$ and $C(u,v)$ be the same as introduced before.

Let graph $\hat{G} = (\hat{V}, \hat{E}, \hat{w}_E)$ be defined as follows:

$$\hat{V} = \{u | u \in V \wedge C(u) > n/2\},$$
$$\hat{E} = \{(u,v) | u, v \in \hat{V}\},$$
$$\hat{w}_E(u,v) = \text{median}\{w_{E_i}^{G_i}(u,v) | i = 1, \ldots, n\}.$$

Then it can be proven that graph $\hat{G}$ is a median of sequence $S$ under $d_2$ [33].

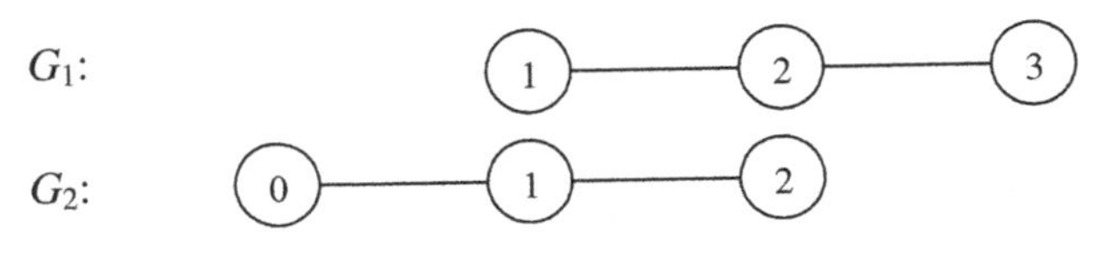


Fig. 1. Two graphs $(G_1, G_2)$.

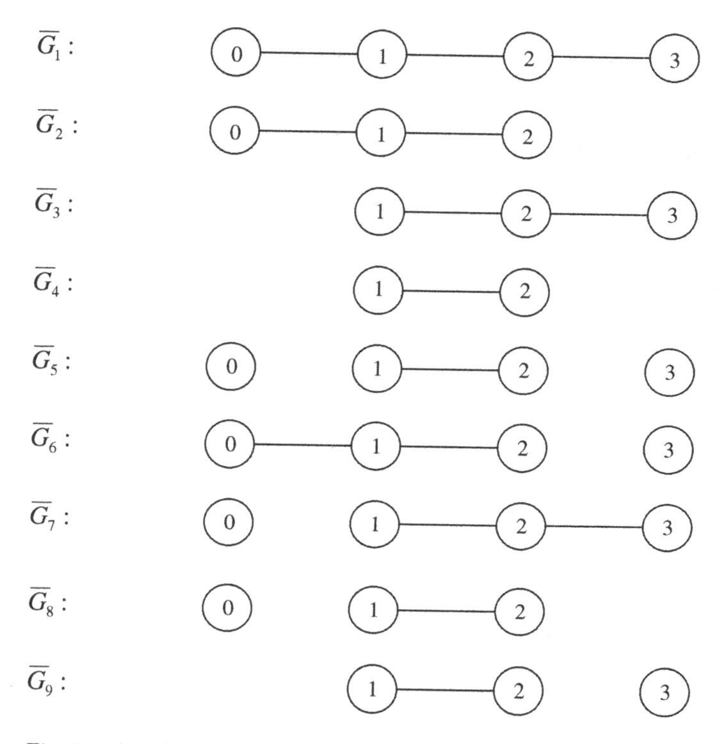


Fig. 2. All possible medians of the sequence $(G_1, G_2)$ from Fig. 1.

Comparing the median graph construction procedure for $d_1$ with the one for $d_2$, we notice that the former is a special case of the latter, constraining edge weights to assume only binary values. Edge weight zero (or one) indicates the absence (or presence) of an edge. Including an edge $(u, v)$ in the median graph $\bar{G}$ because it occurs in more than $n/2$ of the given graphs is equivalent to labeling that edge in $\hat{G}$ with the median of the weights assigned to it in the given graphs.

The median of a set of numbers according to Eq. (5.2) is unique. Hence when constructing a median graph under graph distance measure $d_2$, there will be no ambiguity in edge weights. But the non-uniqueness of the existence of vertices observed for graph distance measure $d_1$ still exists under $d_2$. The convergence property holds also for $d_2$, because for any sequence of edge weights $(x_1, \ldots, x_k)$ one has:

$$\text{median}(x_1, \ldots, x_k) = \text{median}(x_1, \ldots, x_k, \text{median}(x_1, \ldots, x_k)).$$

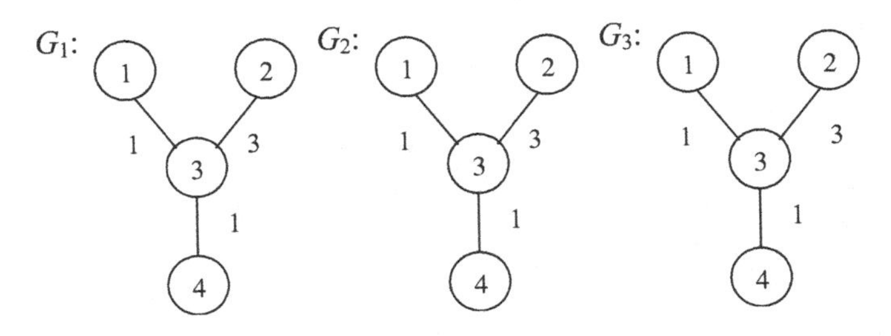


Fig. 3. Three graphs $(G_1, G_2, G_3)$.

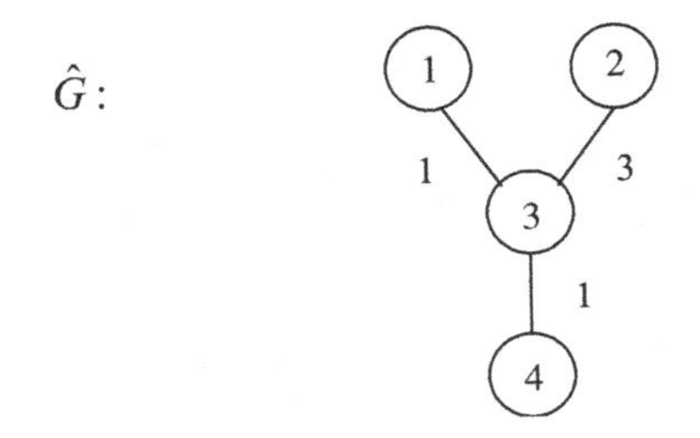


Fig. 4. The median of $(G_1, G_2, G_3)$ in Fig. 3 using $d_2$.

We conclude this section with an example of median graph under graph distance $d_2$. Three different graphs $G_1$, $G_2$, and $G_3$ are shown in Figure 3. Their median is unique and is displayed in Figure 4.

## 6. Median Graphs and Abnormal Change Detection in Sequences of Graphs

The measures defined in Sections 3 and 4 can be applied to consecutive graphs in a time series of graphs $(G_1, \ldots, G_n)$ to detect abnormal change. That is, values of $d(G_{i-1}, G_i)$ are computed for $i = 2, \ldots, n$, and the change from time $i-1$ to $i$ is classified abnormal if $d(G_{i-1}, G_i)$ is larger than a certain threshold. However, it can be argued that measuring network change only between consecutive points of time is potentially vulnerable to noise, i.e., the random appearance or disappearance of some vertices together with some random fluctuation of the edge weights may lead to a graph distance larger than the chosen threshold, though these changes are not really significant.

One expects a more robust change detection procedure is obtained through the use of median graphs. In statistical signal processing the median filter is widely used for removing impulsive noise. A median filter is computed by sliding a window of length $L$ over data values in a time series. At each step the output to this process is the median of values within the window. This process is also termed the *running median* [35]. In the following we discuss four different approaches to abnormal change detection that utilise median filters. All these approaches assume that a time series of graphs $(G_1, \dots, G_n, G_{n+1}, \dots)$ is given. The median graph of a subsequence of these graphs can be computed using either graph distance measure $d_1$ or $d_2$.

### 6.1. *Median vs. Single Graph, Adjacent in Time (msa)*

Given the time series of graphs, we compute the median graph in a window of length $L$, where $L$ is a parameter that is to be specified by the user dependent on the underlying application. Let $\tilde{G}_n$ be the median of the sequence $(G_{n-L+1}, \dots, G_n)$. Then $d(\tilde{G}_n, G_{n+1})$ can be used to measure abnormal change. We classify the change between $G_n$ and $G_{n+1}$ as *abnormal* if $d(\tilde{G}_n, G_{n+1})$ is larger than some threshold.

Increased robustness can be expected if we take the average deviation $\varphi$, of graphs $(G_{n-L+1}, \dots, G_n)$ into account. We compute

$$\varphi = \frac{1}{L} \sum_{i=n-L+1}^{n} d(\tilde{G}_n, G_i) \tag{6.1}$$

and classify the change between $G_n$ and $G_{n+1}$ abnormal if

$$d(\tilde{G}_n, G_{n+1}) \geq \alpha\varphi \tag{6.2}$$

where $\alpha$ is a parameter that needs to be determined from examples of normal and abnormal change. Note that the median $\tilde{G}_n$ is, by definition, a graph that minimises $\varphi$ in Eq. (6.1).

In Section 5 it was pointed out that $\tilde{G}_n$ is not necessarily unique. If several instances of $\tilde{G}_{n_1}, \dots, \tilde{G}_{n_1}$ of $\tilde{G}_n$ exist, one can apply Eqs. (6.1) and (6.2) on all of them. This will result in a series of values $\varphi_1, \dots, \varphi_1$ and a series of values $d(\tilde{G}_{n_1}, G_{n+1}), \dots, d(\tilde{G}_{n_1}, G_{n+1})$. Under a conservative scheme, an abnormal change will be reported if $d(\tilde{G}_{n_1}, G_{n-1}) \geq \alpha\varphi_1 \wedge \cdots \wedge d(\tilde{G}_{n_t}, G_{n+1}) \geq \alpha\varphi_t$.

By contrast a more sensitive change detector is obtained if a change is reported as soon as there exists at least one $i$ for which

$$d(\tilde{G}_{n_i}, G_{n+1}) \geq \alpha\varphi_i, \quad 1 \leq i \leq t.$$

### 6.2. *Median vs. Median Graph, Adjacent in Time* (mma)

Here we compute two median graphs, $\tilde{G}_1$ and $\tilde{G}_2$, in windows of length $L_1$ and $L_2$, respectively, i.e., $\tilde{G}_1$ is the median of the sequence $(G_{n-L_1+1}, \ldots, G_n)$ and $\tilde{G}_2$ is the median of the sequence $(G_{n+1}, \ldots, G_{n+L_2})$. We measure now the abnormal change between time $n$ and $n+1$ by means of $d(\tilde{G}_1, \tilde{G}_2)$. That is, we compute $\varphi_1$ and $\varphi_2$ for each of the two windows using Eq. (6.1) and classify the change from $G_n$ to $G_{n+1}$ as abnormal if

$$d(\tilde{G}_1, \tilde{G}_2) \geq \alpha \left[ \frac{L_1 \varphi_1 + L_2 \varphi_2}{L_1 + L_2} \right].$$

Measure `mma` can be expected even more robust against noise and outliers than measure `msa`. If the considered median graphs are not unique, similar techniques (discussed for measure `msa`) can be applied.

### 6.3. *Median vs. Single Graph, Distant in Time* (msd)

If graph changes are evolving rather slowly over time, it may be better not to compare two consecutive graphs, $G_n$ and $G_{n+1}$, with each other, but $G_n$ and $G_{n+l}$, where $l > 1$. Instead of `msa`, as proposed above, we use $d(\tilde{G}_n, G_{n+1})$ as a measure of change between $G_n$ and $G_{n+l}$, where $l$ is a parameter defined by the user and is dependent on the underlying application.

### 6.4. *Median vs. Median Graph, Distant in Time* (mmd)

This measure is a combination of the measures mma and msd. We use $\tilde{G}_1$ as defined for `mma`, and let $\tilde{G}_2 = \text{median}(G_{n+l+1}, \ldots, G_{n+l+L_2})$. Then $d(\tilde{G}_1, \tilde{G}_2)$ can serve as a measure of change between time $n$ and $n+l+1$. Obviously, Eqs. (6.1) and (6.2) can be adapted to `msd` and `mmd` similarly to the way they are adapted to `mma`.

## 7. Application to Computer Network Monitoring

### 7.1. *Problem Description*

In managing large enterprise data networks, the ability to measure network changes in order to detect abnormal trends is an important performance monitoring function [36]. The early detection of abnormal network events and trends can provide advance warning of possible fault conditions

[37], or at least assist with identifying the causes and locations of known problems.

Network performance monitoring typically uses statistical techniques to analyse variations in traffic distribution [38,39], or changes in topology [40]. Visualisation techniques are also widely used to monitor changes in network performance [41]. To complement these approaches, specific measures of change at the network level in both logical connectivity and traffic variations are useful in highlighting when and where abnormal events may occur in the network [42]. Using these measures, other network management tools may then be focused on problem regions of the network for more detailed analysis.

In the previous sections, various graph similarity measures are introduced. The aim of the study described in the present section is to identify whether using these techniques, significant changes in logical connectivity or traffic distributions can be observed between large groups of users communicating over a wide area data network. This data network interconnects some 120,000 users around Australia. For the purposes of this study, a network management probe was attached to a physical link on the wide area network backbone and collected traffic statistics of all data traffic operating over the link. From this information a logical network of users communicating over the physical link is constructed.

Communications between user groups (business domains) within the logical network over any one day is represented as a directed graph. Edge direction indicates the direction of traffic transmitted between two adjacent nodes (business domains) in the network, with edge-weight indicating the amount of traffic carried. A subsequent graph can then describe communications within the same network for the following day. This second graph can then be compared with the original graph, using a measure of graph distance between the two graphs, to indicate the degree of change occurring in the logical network. The more dissimilar the graphs, the greater the graph distance value. By continuing network observations over subsequent days, the graph distance scores provide a trend of the logical network's relative dynamic behaviour as it evolves over time.

In the study described in this section, log files were collected continuously over the period 9 July 1999 to 24 December 1999. Weekends, public holidays and days where probe data was unavailable were removed to produce a final set of 102 log files representing the successive business days' traffic. The graph distance measures examined in this paper produce a distance score indicating the dissimilarity between two given graphs.

Successive graphs derived from the 102 log files of the network data set are compared using the various graph distance measures to produce a set of distance scores representing the change experienced in the network from one day to the next.

### 7.2. *Experimental Results*

Figures 5 and 6 represent the outcomes of connectivity and traffic (weighted graph) spectral distances, respectively, as introduced in Section 3, applied to the time series of graphs derived from the network data. In spite of the less intuitive interpretation of these graph distance measures, there exists reasonable correlation with the peaks of other distance measures and far less sensitivity to daily variations using this approach (see below).

Figure 7 shows results for edit distance applied to consecutive graphs of the time series for the topology only measure $d_1$. This measure produces three significant peaks (on days 25, 65 and 90). The figure also shows several secondary peaks that may also indicate events of potential interest. Also, there is significant minor fluctuation throughout the whole data set

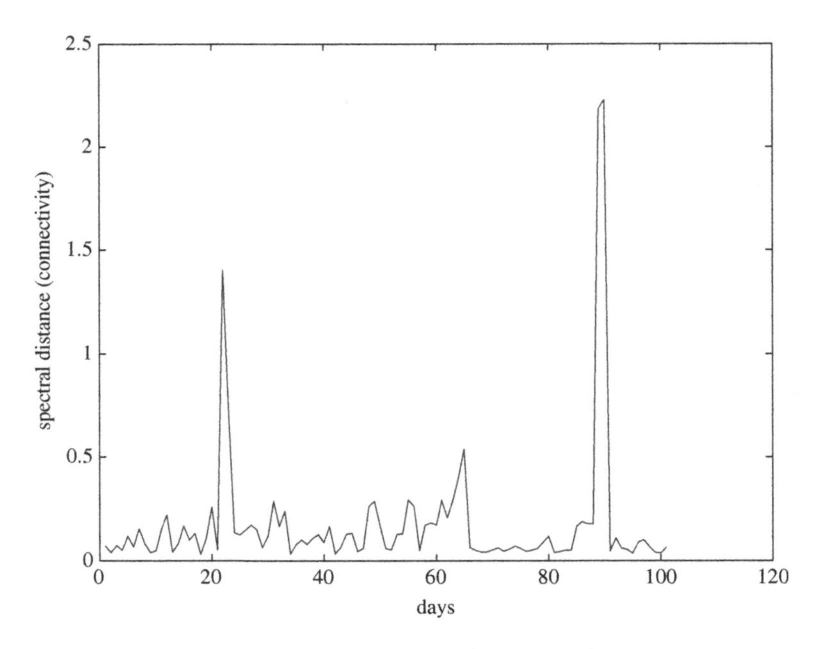


Fig. 5. Special distance (connectiviy).

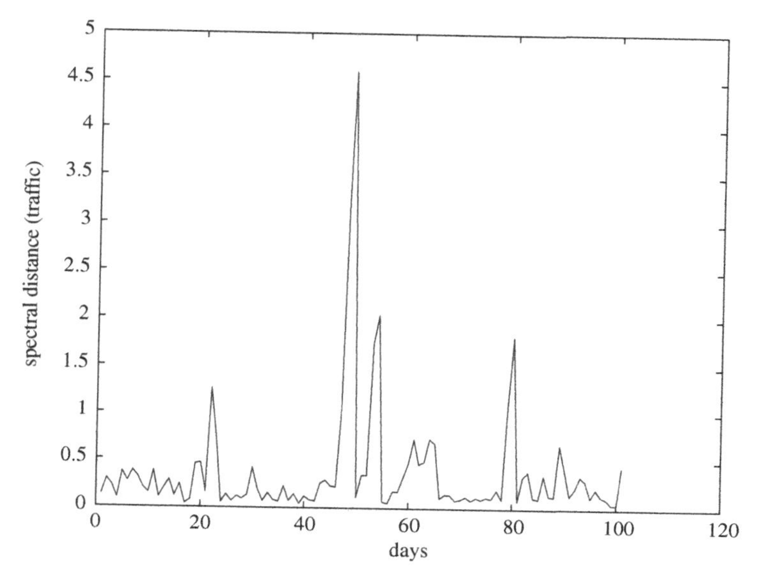


Fig. 6. Special distance (traffic).

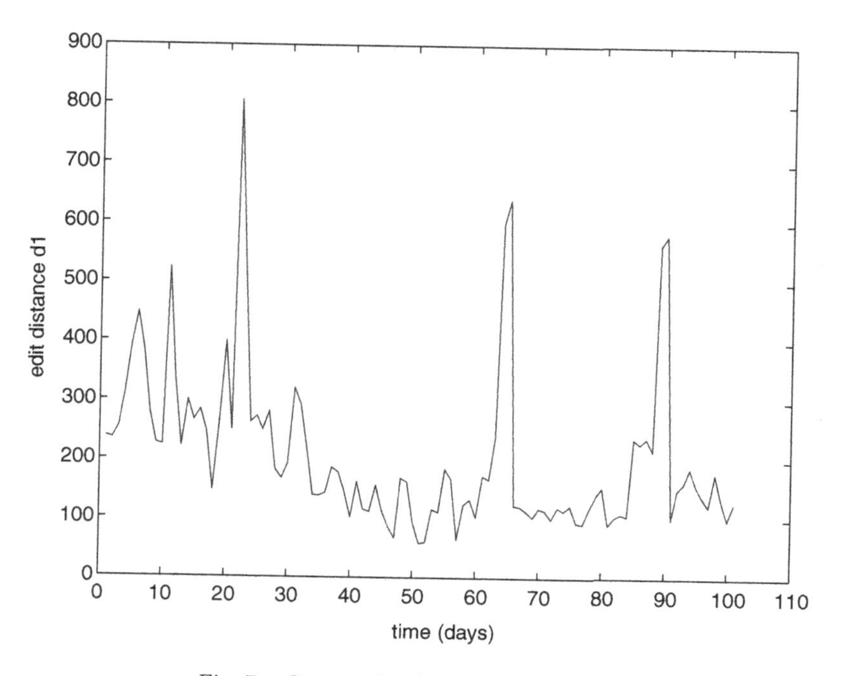


Fig. 7. Consecutive day using measure $d_1$.

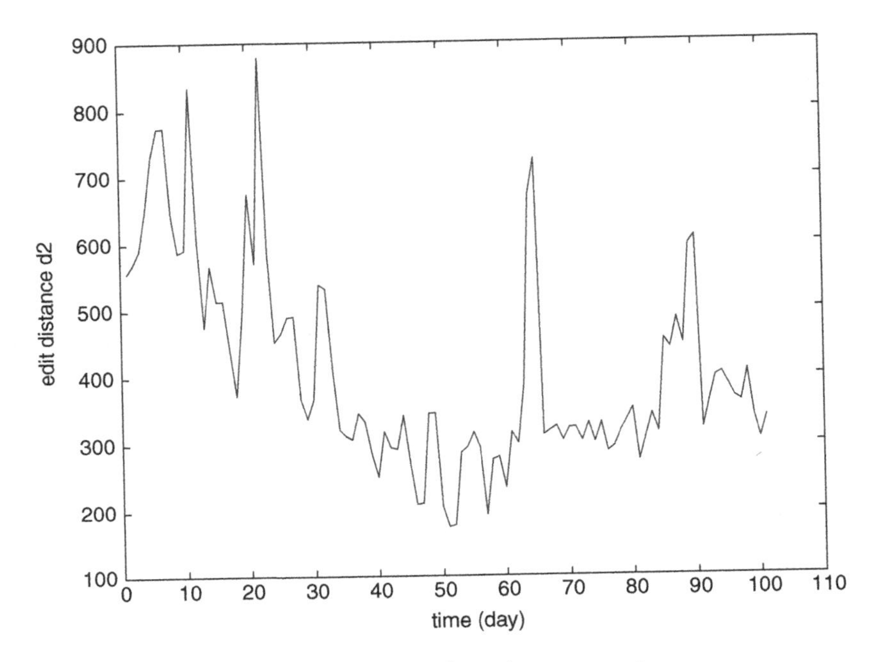


Fig. 8. Consecutive day using measure $d_2$.

appearing as background noise. Figure 8 shows the results for distance measure $d_2$ being applied to consecutive graphs in the time series. Whilst this measure considers both topology and traffic, it does not appear to provide any additional indicators of change additional to those found using the topology only measure. The main difference with this measure is that it has increased the amplitude of the first two peaks so that they now appear as major peaks. This suggests that these peaks were a result of network change consisting of large change in edge weights.

The findings for this technique indicate that the application of measures to consecutive graphs in a time series is most suited for detecting daily fluctuations in network behaviour. This is especially useful for identifying outliers in a time series of graphs. Ideally, it would be desirable that each point of significant change be validated against known performance anomalies. Unfortunately, information generated by the network management system of the network under observation is not sufficient for validation purposes, because all data used in the experiments were collected from a network under real operating conditions. That is, the data are unlabelled as there is no 'super-instance' that would know whether a network change that has occurred is normal or abnormal. This shortfall of useful information for validation of network fault and performance anomalies is common [43]. Visualisation tools have been used to assist with validating some of the major changes by observing the network before and after a significant change event. More work needs to be done in producing a reliable, validated data set.

A number of additional experiments using median graphs have been reported in [44]. In these experiments it was demonstrated that using the median graph, computed over a running window of the considered time series, rather than individual graphs has a smoothing effect, which reduces the influence of outliers.

## 8. Conclusion

Graphs are one of the most popular data structures in computer science and related fields. This chapter has examined a special class of graphs that are characterised by the existence of unique node labels. This property greatly reduces the computational complexity of many graph operations and allows a user to deal with large sets of graphs, each consisting of many nodes and edges. Furthermore, a number of graph distance measures have been analysed. They are based on eigenvalues from spectral graph theory and graph edit distance. A novel concept that can be used to measure the similarity of graphs is median graph. The median of a set, or a sequence, $S$, of graphs is a graph that minimises the average edit distance to all members in $S$. Thus the median can be regarded as the best single representative of a given set or sequence of graphs.

Abnormal events in a sequence of graphs representing elements of a time dependent process can be detected by computing the distance of consecutive pairs of graphs in the sequence. If the distance is larger than a threshold than it can be concluded that an abnormal event has occurred. Alternatively, to make the abnormal change detection procedure more robust against noise and outliers, one can compute the median of the time series of graphs over a window of a given span, and compare it to an individual graph, or the median of a sequence of graphs, following that window.

None of the formal concepts introduced in this paper is geared towards a particular application. However, to demonstrate their usefulness in practice, an application in computer network monitoring was studied in this chapter. Communication transactions, collected by a network management system between logical nodes occurring over periodic time intervals are represented as a series of weighted graphs. The graph distance measures introduced in this chapter are used to assess the changes in communication between user groups over successive time intervals. Using these data in a number of experiments, the usefulness of the proposed concepts could be established. As a practical implication, the proposed methods have the potential to relive human network operators from the need to continually monitor a network, if connectivity and traffic patterns can be shown to be similar to the activity over previous time periods. Alternatively, if traffic volumes are seen to abruptly increase or decrease over the physical link, network operators are able to more readily identify the individual user groups contributing to this change in aggregated traffic. Future work will concentrate on enhancing the present tools for computer network monitoring and identifying other application areas for the proposed formal concepts.

## References

1. Rouvray, D.H. and Balaban, A.T. (1979). Chemical Applications of Graph Theory. In R.J. Wilson and L.W. Beineke, eds. *Applications of Graph Theory*, pp. 177–221. Academic Press.
2. Baxter, K. and Glasgow, J. (2000). Protein Structure Determination: Combinnig Inexact Graph Matching and Deformable Templates. *Proc. Vision Interface 2000*, pp. 179–186.
3. Rodgers, P.J. and King, P.J.H. (1997). A Graph–Rewriting Visual Language for Database Programming. *Journal of Visual Languages and Computing*, **8**, 641–674.
4. Shearer, K., Bunke, H., and Venkatesh, S. Video Indexing and Similarity Retrieval by Largest Common Subgraph Detection Using Decision Trees. *Pattern Recognition*, **34**, 1075–1091.
5. Börner, K., Pippig, E., Tammer, E., and Coulon, C. (1996). Structural Similarity and Adaption. In I. Smith and B. Faltings, eds. *Advances in Case-based Reasoning*, pp. 58–75. LNCS 1168, Springer.
6. Fisher, D.H. (1990). Knowledge Acquisition Via Incremental Conceptual Clustering. In J.W. Shavlik and T.G. Dietterich, eds. *Readings in Machine Learning*, pp. 267–283. Morgan Kaufmann.
7. Sanders, K., Kettler, B., and Hendler, J. (1997). The Case for Graph-Structured Representations. In D. Leake and E. Plaza, eds. *Case-Base Reasoning Research and Development*, **1266** of *Lecture Notes in Computer Science*, pp. 245–254. Springer.
8. Ehrig, H. (1992). Introduction to Graph Grammars with Applications to Semantic Networks. *Computers and Mathematics with Applications*, **23**, 557–572.

9. Djoko, S., Cook, D.J., and Holder, L.B. (1997). An Empirical Study of Domain Knowledge and its Benefits to Substructure Discovery. *IEEE Trans. Knowledge and Data Engineering*, **9**, 575–586.
10. Lu, S.W., Ren, Y., and Suen, C.Y. (1991). Hierarchical Attributed Graph Representation and Recognition of Handwritten Chinese Characters. *Pattern Recognition*, **24**, 617–632.
11. Llados, J., Marti, E., and Villanueva, J.J. (2001). Symbol Recognition by Error-Tolerant Subgraph Matching Between Region Adjacency Graphs. *IEEE Trans. Pattern Analysis and Machine Intelligence*, **23**, 1144–1151.
12. Shokonfandeh, A. and Dickinson, S. (2001). A Unified Framework for Indexing and Matching Hierarchical Shape Structures. In C. Arcelli, L. Cordella, and G. Sanniti di Baja, eds. *Visual Form 2001*, pp. 67–84. Springer Verlag, LNCS 2059.
13. Kittler, J. and Ahmadyfard, A. (2001). On Matching Algorithms for the Recogniton of Objects in Cluttered Background. In C. Arcelli, L. Cordella, and G. Sanniti di Baja, Eds. *Visual Form 2001*, pp. 51–66. Springer Verlag, LNCS 2059.
14. Mitchell, T.M. (1997). *Machine Learning.* Mc Graw-Hill.
15. Messmer, B.T. and Bunke, H. (1998). A New Algorithm for Error-Tolerant Subgraph Isomorphism Detection. *IEEE Trans. Pattern Anal. & Machine Intell.*, **20**, 493–504.
16. Fayyad, U., Piatetsky-Sharpiro, G., Smyth, P., and Uthurusamy, R. (1996). *Advances in Knowledge Discovery and Data Mining.* AAAI/MIT Press.
17. Kandel, A., Last, M., and Bunke, H. (2001). *Data Mining and Computational Intelligence.* Physica Verlag.
18. Chen, M.S., Han, J., and Yu, P.S. (1996). Data Mining: An Overview from a Database Perspective. *IEEE Trans. Data & Knowledge Engineering*, **8**, 866–883.
19. Bunke, H., Kraetzl, M., Shoubridge, P. and Wallis, W.D. Measuring Abnormal Change in Large Data Networks. *Submitted for publications.*
20. Strogaz, S.H. (2001). Exploring Complex Networks. *Nature*, **40**, 268–276.
21. Kuhn, K.W. (1999). Molecular Interaction Map of the Mammalian Cell Cycle Control and DNA Repair System. *Mol. Biol. Cell*, **10**, pp. 2703–2734.
22. Newman, M.E.J. (2001). The Structure of Scientific Collaboration Networks. *Proc. Nat. Acad. Science USA*, **98**, 404–409.
23. West, D.B. (1996). *Introduction to Graph Theory.* Prentice Hall, New Jersey.
24. Cvetkovic, D.M., Doob, M., and Sachs, H. (1980). *Spectra of Graphs.* Academic Press, New York.
25. Mohar, B. (1992). Laplace Eigenvalues of Graphs — A Survey. *Discrete Math.*, **109**, 171–183.
26. Sarkar, S. and Boyer, K.L. (1998). Quantitative Measures of Change Based on Feature Organization: Eigenvalues and Eigenvectors. *Computer Vision and Image Understanding*, **71**, 110–136.
27. Umeyama, S. (1988). An Eigendecomposition Approach to Weighted Graph Matching Problems. *IEEE Trans. Pattern Anal.& Machine Intell.*, **10**(5), 695–703.

28. Guattery, S. and Miller, G.L. (1998). On the Quality of Spectral Separators. *SIAM J. Matrix Anal. Appl.*, **19**(3), 701–719.
29. Chung, F.R.K. (1997). *Spectral Graph Theory*. CBMS, Regional Conference Series in Mathematics. American Mathematical Society.
30. Fiedler, M. (1973). Algebraic Connectivity of Graphs. *Czech. Math. J.*, **23**(98), 298–305.
31. Sanfeliu, A. and Fu, K.S. (1983). A Distance Measure Between Attributed Relational Graphs for Pattern Recognition. *IEEE Transactions on Systems Management and Cybernetics*, **13**(3), 353–362.
32. Bunke, H. and Messmer, B.T. (1997). Recent Advances in Graph Matching. *International J. Pattern Recognition and Artificial Intelligence*, **11**(1), 169–203.
33. Dickinson, P., Bunke, H., Dadej, A., and Kraetzl, M. (2001). Application of Median Graphs in Detection of Abnomalous Change in Communication Networks. In *Proc. World Multiconference on Systemics, Cybernetics and Informatics*, **5**, pp. 194–197. Orlando, US.
34. Jiang, X. and Bunke, H. On median Graphs: Properties, Algorithms, and Applications. *IEEE Trans. Pattern Anal. and Machine Intell.*, **23**(10), 1144–1151.
35. Astola, J.T. and Campbell, T.G. (1989). On Computation of the Running Median. *IEEE Trans. Acoustics, Speech, and Signal Processing*, **37**(4), 572–574.
36. Boutaba, R., Guemhioui, K.E., and Dini, P. (1997). An Outlook on Intranet Management. *IEEE Communications Magazine*, pp. 92–99.
37. Thottan, M. and Ji, C. (1998). Proactive Anomaly Detection Using Distributed Intelligent Agents. *IEEE Network*, **12**(5), 21–27.
38. Higginbottom, G.N. (1998). *Performance Evaluation of Communication Networks*. Artech House, Massachusetts.
39. Jerkins, J.L. and Wang, J.L. (1998). A Close Look at Traffic Measurements from Packet Networks. In *IEEE GLOBECOM 98, Sydney*, **4**, 2405–2411.
40. White, C.C., Sykes, E.A., and Morrow, J.A. (1995). An Analytical Approach to the Dynamic Topology Problem. *Telecommunication Systems*, **3**, 397–413.
41. Becker, R.A., Eick, S.G., and Wilks, A.R. (1995). Visualizing Network Data. *IEEE Trans. Visualization and Computer Graphics*, **1**(1), 16–21.
42. Shoubridge, P., Kraetzl, M., and Ray, D. (1999) Detection of Abnormal Change in Dynamic Networks. In *IEEE Information, Decision and Control, IDC '99 conference*, pp. 557–562. Adelaide, Australia.
43. Hood, C.S. and Ji, C. (1997). Beyond Thresholds: An Alternative Method for Extracting Information from Network Measurements. In *IEEE Global Telecommunications Conference, GLOBECOM '97*, **1**, 487–491.
44. Dickinson, P., Bunke, H., Dadej, A., and Kraetzl, M. (2001). Application of Median Graphs in Detection of Anomalous Change in Time Series of Networks. Submitted for publication.

# CHAPTER 7

# BOOSTING INTERVAL-BASED LITERALS: VARIABLE LENGTH AND EARLY CLASSIFICATION*

Carlos J. Alonso González
*Dpto. de Informática, Universidad de Valladolid, Spain*
*Grupo de Sistemas Inteligentes*
E-mail: calonso@infor.uva.es

Juan J. Rodríguez Diez
*Lenguajes y Sistemas Informáticos, Universidad de Burgos, Spain*
*Grupo de Sistemas Inteligentes*
E-mail: jjrodriguez@ubu.es

This work presents a system for supervised time series classification, capable of learning from series of different length and able of providing a classification when only part of the series are presented to the classifier. The induced classifiers consist of a linear combination of literals, obtained by *boosting* base classifiers that contain only one literal. Nevertheless, these literals are specifically designed for the task at hand and they test properties of fragments of the time series on temporal intervals. The method had already been developed for fixed length time series. This work exploits the symbolic nature of the classifier to add it two new features. First, the system has been slightly modified in order that it is now able to learn directly from variable length time series. Second, the classifier can be used to identify partial time series. This "early classification" is essential in some task, like on line supervision or diagnosis, where it is necessary to give an alarm signal as soon as possible. Several experiments on different data test are presented, which illustrate that the proposed method is highly competitive with previous approaches in terms of classification accuracy.

*Keywords*: Interval based literal; boosting; time series classification; machine learning.

*This work has been supported by the Spanish CYCIT project TAP 99–0344 and the "Junta de Castilla y León" project VA101/01.

## 1. Introduction

Multivariate time series classification is useful in those classification tasks where time is an important dimension. Instances of these kind of tasks may be found in very different domains, for example analysis of biomedical signals [Kubat *et al.* (1998)], diagnosis of continuous dynamic systems [Alonso González and Rodríguez Diez (1999)] or data mining in temporal databases [Berndt and Clifford (1996)]. Time series classification may be addressed like an static classification problem, extracting features of the series through some kind of preprocessing, and using some conventional machine learning method. However, this approach has several drawbacks [Kadous (1999)]: the preprocessing techniques are usually *ad hoc* and domain specific, and the descriptions obtained using these features can be hard to understand. The design of specific machine learning methods for the induction of time series classifiers allows for the construction of more comprehensible classifiers in a more efficient way because, firstly, they may manage comprehensible temporal concepts difficult to capture by the preprocessing technique — for instance the concept of permanence in a region for certain amount of time — and secondly, there are several heuristics applicable to temporal domains that preprocessing methods fails to exploit.

The method for learning time series classifiers that we propose in this work is based on literals over temporal intervals (such as increases or always in region) and boosting (a method for the generation of ensembles of classifiers from base or weak classifiers) [Schapire (1999)] and was first introduced in [Rodríguez *et al.* (2000)]. The input for this learning task consist of a set of examples and associated class labels, where each example consists of one or more time series. Although the series are often referred to as variables, since they vary over time, form a machine learning point of view, each point of each series is an attribute of the example. The output of the learning task is a weighted combination of literals, reflecting the fact that the based classifiers consist of clauses with only one literal. These base classifiers are inspired by the good results of works using ensembles of very simple classifiers [Schapire (1999)], sometimes named *stumps*.

Although the method has already been tested over several data set [Rodríguez *et al.* (2001)] providing very accurate classifiers, it imposes two restrictions that limits its application to real problems. On the one hand, it requires that all the time series were of the same length, which is not the case in every task, as the Auslan example (Australian sign language, see Section 6.4) [Kadous (1999)] illustrates. On the other hand, it requires as an input the completed time series to be classified. This may be a severe drawback if the classifiers are to be use on line in a dynamic environment and a classification is needed as soon as possible. This work shows how the original method can be extended to cope with these weaknesses.

Regarding variable length time series, it is always possible to preprocess the data set in order to obtain a new one with fixed length series. Nevertheless, this cannot be considered as a generic solution, because, usually, the own length of the series provides essential information for its classification. The method can now be used with variable length series because the literals are allowed to abstain (that is, the result of their evaluation can be true, false or an abstention) if the series is not long enough to evaluate some literal. A variant of boosting that can work with base classifiers with abstentions is used.

Regarding the classification of partial examples, or early classification, the system has been modified to add it the capacity of assigning a preliminary classification to an incomplete example. This feature is crucial when the examples to classify are being generated dynamically and the time necessary to generate an example is long. For instance, consider a supervision task of a dynamic system. In this case, the example to classify is the current state, considering the historical values of the variables. For this problem, it is necessary to indicate a possible problem as soon as possible. In order to confirm the problem, it will be necessary to wait and see how the variables evolve. Hence, it is necessary to obtain classifications using as input series of different lengths. Again, the capability of the literals to abstain allows tackling this problem.

The rest of the chapter is organized as follows. Section 2 is a brief introduction to boosting, suited to our method. The base classifiers, interval literals, are described in Section 3. Sections 4 and 5 deal, respectively, with variable length series and early classification. Section 6 presents experimental results. Finally, we give some concluding remarks in Section 7.

## 2. Boosting

At present, an active research topic is the use of *ensembles* of classifiers. They are obtained by generating and combining base classifiers, constructed using other machine learning methods. The target of these ensembles is to increase the accuracy with respect to the base classifiers.

One of the most popular methods for creating ensembles is boosting [Schapire (1999)], a family of methods, of which AdaBoost is the most prominent member. They work by assigning a weight to each example. Initially, all the examples have the same weight. In each iteration a *base* (also named *weak*) classifier is constructed, according to the distribution of weights. Afterwards, the weight of each example is readjusted, based on the correctness of the class assigned to the example by the base classifier. The final result is obtained by weighted votes of the base classifiers.

ADABOOST is only for binary problems, but there are several methods of extending AdaBoost to the multiclass case. Figure 1 shows AdaBoost.MH[1] [Schapire and Singer (1998)]. Each instance $x_i$ belongs to a domain $X$ and has an associated label $y_i$, which belongs to a finite label space $Y$. AdaBoost.MH associates a weight to each combination of examples and labels. The base learner generates a base hypothesis $h_t$, according to the weights. A real value, $\alpha_t$, the weight of the base classifier, is selected. Then, the weights are readjusted. For $y, l \in Y, y[l]$ is defined as

$$y[l] = \begin{cases} +1 & \text{if } l = y, \\ -1 & \text{if } l \neq y. \end{cases}$$

Two open questions are how to select $\alpha_t$ and how to train the weak learner. The first question is addressed in [Schapire and Singer (1998)]. For two class problems, If the base classifier returns a value in $\{-1, +1\}$, then

---

Given $(x_1, y_1), \ldots, (x_m, y_m)$ where $x_i \in X, y_i \in Y$
Initialize $D_1(i, l) = 1/(mk)$
For $t = 1, \ldots, T$:

- Train weak learner using distribution $D_t$
- Get weak hypothesis $h_t : X \times Y \longrightarrow \mathbf{R}$
- Choose $\alpha_t \in \mathbf{R}$
- Update

  $$D_{t+1}(i, l) = D_t(i, l) \ \exp(-\alpha_t y_i[l] \ h_t(x_i, l))/Z_t$$

  where $Z_t$ is a normalization factor (chosen so that $D_{t+1}$ will be a distribution)

Output the final hypothesis

---

Fig. 1. AdaBoost.MH [Schapire and Singer (1998)].

[1] Although ADABOOST.MH also considers multilabel problems (an example can be simultaneously of several classes), the version presented here does not include this case.

the best is

$$\alpha = \frac{1}{2}\ln\left(\frac{W_+}{W_-}\right).$$

Where $W_+$ and $W_-$ are, respectively, the sum of the weights of the examples well and bad classified.

For the last question, how to train the base learner, it must be a multi-class learner, and we want to use binary learners (only one literal). Then, for each iteration, we train the weak learner using a binary problem: one class (selected randomly) against the others. The output of this weak learner, $h_t^B(x)$ is binary. On the other hand, we do not generate a unique $\alpha_t$, but for each class, $l$, an $\alpha_{tl}$ is selected. They are selected considering how good is the weak learner for discriminating between the class $l$ and the rest of classes. This is a binary problem so the selection of the values can be done as indicated in [Schapire and Singer (1998)]. Now, we can define $h_t(x,l) = \alpha_{tl}h_t^B(x)$ and $\alpha_t = 1$ and we use AdaBoost.MH.

## 3. Interval Based Literals

Figure 2 shows a classification of the predicates used to describe the series.

Point based predicates use only one point of the series:

- point_le(Example, Variable, Point, Threshold) it is true if, for the Example, the value of the Variable at the Point is less or equal than Threshold.

Note that a learner that only uses this predicate is equivalent to an attribute-value learning algorithm. This predicate is introduced to test the results obtained with boosting without using interval based predicates.

Two kinds of interval predicates are used: relative and region based. Relative predicates consider the differences between the values in the interval. Region based predicates are based on the presence of the values of a variable in a region during an interval. This section only introduces the predicates [Rodríguez *et al.* (2001)] gives a more detailed description, including how to select them efficiently.

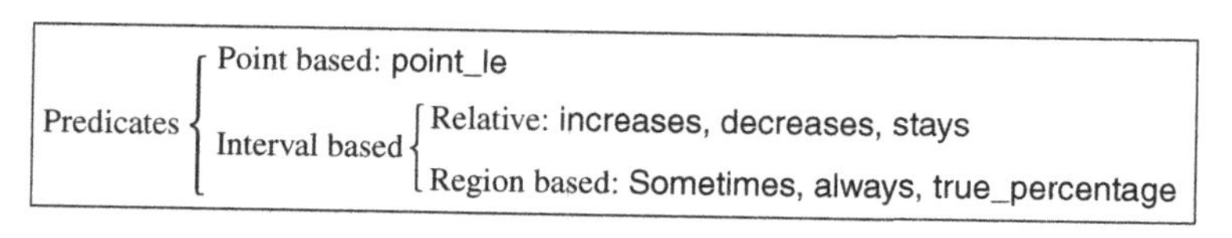


Fig. 2. Classification of the predicates.

### 3.1. *Relative Predicates*

A natural way of describing series is to indicate when they increase, decrease or stay. These predicates deal with these concepts:

- increases(Example, Variable, Beginning, End, Value). It is true, for the Example, if the difference between the values of the Variable for End and Beginning is greater or equal than Value.
- decreases(Example, Variable, Beginning, End, Value).
- stays(Example, Variable, Beginning, End, Value). It is true, for the Example, if the range of values of the Variable in the interval is less or equal than Value.

Frequently, series are noisy and, hence, a strict definition of increases and decreases in an interval i.e., the relation holds for all the points in the interval, is not useful. It is possible to filter the series prior to the learning process, but we believe that a system for time series classification must not rely on the assumption that the data is clean. For these two predicates we consider what happens only in the extremes of the interval. The parameter Value is necessary for indicating the amount of change.

For the predicate stays it is neither useful to use a strict definition. In this case all the points in the interval are considered. The parameter Value is used to indicate the maximum allowed difference between the values in the interval.

### 3.2. *Region Based Predicates*

The selection and definition of these predicates is based in the ones used in a visual rule language for dynamic systems [Alonso González and Rodríguez Diez (1999)]. These predicates are:

- always(Example, Variable, Region, Beginning, End). It is true, for the Example, if the Variable is always in this Region in the interval between Beginning and End.
- sometimes(Example, Variable, Region, Beginning, End).
- true_percentage(Example, Variable, Region, Beginning, End, Percentage). It is true, for the Example, if the percentage of the time between Beginning and End where the variable is in Region is greater or equal to Percentage.

Once that it is decided to work with temporal intervals, the use and definition of the predicates always and sometimes is natural, due to the fact that they are the extension of the conjunction and disjunction to intervals.

Since one appears too demanding and the other too flexible, a third one has been introduced, true_percentage. It is a "relaxed always" (or a "restricted sometimes"). The additional parameter indicates the degree of flexibility (or restriction).

**Regions.** The regions that appear in the previous predicates are intervals in the domain of values of the variable. In some cases the definitions of these regions can be obtained from an expert, as background knowledge. Otherwise, they can be obtained with a discretization preprocess, which obtains $r$ disjoint, consecutive intervals. The regions considered are these $r$ intervals (equality tests) and others formed by the union of the intervals $1, \ldots, i$ (less or equal tests).

The reasons for fixing the regions before the classifier induction, instead of obtaining them while inducing, are efficiency and comprehensibility. The literals are easier to understand if the regions are few, fixed and not overlapped.

### 3.3. *Classifier Example*

Table 1 shows a classifier. It was obtained from the data set CBF (Section 6.1). This classifier is composed by 10 base classifiers. The first column shows the literal. For each class in the data set there is another column, with the weight associated to the literal for that class.

In order to classify a new example, a weight is calculated for each class, and then the example is assigned to the class with greater weight. Initially, the weight of each class is 0. For each base classifier, the literal is evaluated. If it is true, then, for each class, its weight is updated adding the weight of

Table 1. Classifier example, for the CBF data set (Section 6.1). For each literal, a weight is associated to every class.

| Literal | Cylinder | Bell | Funnel |
|---|---|---|---|
| true_percentage(E, x, 1_4, 4, 36, 95) | −0.552084 | 3.431576 | −0.543762 |
| not true_percentage(E, x, 1_4, 16, 80, 40) | 1.336955 | 0.297363 | −0.527967 |
| true_percentage(E, x, 3, 49, 113, 25) | −0.783590 | −0.624340 | 1.104166 |
| decreases(E, x, 34, 50, 1.20) | −0.179658 | 0.180668 | 0.771224 |
| not true_percentage(E, x, 4, 14, 78, 15) | 0.899025 | −0.234799 | −0.534271 |
| decreases(E, x, 35, 51, 1.20) | −0.289225 | −0.477789 | 0.832881 |
| decreases(E, x, 32, 64, 0.60) | −0.028495 | −0.462483 | 0.733433 |
| not true_percentage(E, x, 3, 12, 76, 15) | 0.971446 | −0.676715 | −0.248722 |
| true_percentage(E, x, 1_4, 18, 34, 95) | 0.085362 | 3.103075 | −0.351317 |
| true_percentage(E, x, 1_3, 34, 38, 30) | 0.041179 | 2.053417 | 0.134683 |

the class for the literal. If the literal is false, then the weight of the class for the literal is subtracted from the weight of the class.

The first literal in the table "true_percentage(E, x, 1_4, 4, 36, 95)" has associated for each class the weights −0.55, 3.43 and −0.54. This means that if the literal is true (false), it is likely that the class is (not) the second one. This literal is not useful for discriminating between the first and the third class.

## 4. Variable Length Series

Due to the method used to select the literals that forms the base classifiers, the learning system that we have introduced in the preceding sections requires series of equal length. Consequently, to apply it to series of different lengths, it is necessary to preprocess the data, normalizing the length of the series.

There are several methods, more or less complex, that allow normalizing the length of a set of series. These methods, which preprocess the data set, can be adequate for some domains. However, the use of these methods is not a general solution, because they destroy a piece of information that may be useful for the classification: the own length of the series. Therefore, it is important that the learning method can deal with variable length series. Of course, it can still be used with preprocessed data sets of uniform length.

To learn from series of variable length, we have opted for a slight modification of the learning algorithm, that allow a literal to inhibit — or abstain — when there are not enough data for its evaluation. This modification also requires some change on the boosting method. With the basic boosting method, for binary problems, the base classifiers return +1 or −1. Nevertheless, there are variants that use confidence-rated predictions [Schapire and Singer (1998)]. In this case, the base learner returns a real value: the sign indicates the classification and the absolute value the confidence in this prediction. A special case consists in the use of three values: −1, 0 and +1. The value 0 indicates that that base classifier abstains. The value of $\alpha$ is again selected as

$$\alpha = \frac{1}{2}\ln\left(\frac{W_+}{W_-}\right).$$

Until now, a literal could be true or false, because all the series had the same length. If the series are of different lengths, there will be literals with intervals that are after the end of the shortest series. For these cases, the result of the evaluation of the literal could be an abstention, a 0, but due to the nature of the literals it will not always be 0. If the end of the series is before the beginning the interval, the result will be 0. If the end of the series is in the interval, the result depends on the predicate:

- For increases and decreases the result will be always 0, because only the extremes of the interval are considered.
- For stays the result will be $-1$ if the available points are not in the allowed range. In other case the result will be 0.
- For always, the result will be $-1$ if there is some available point in the interval that is out of the region. In other case the result will be 0.
- For sometimes, the result will be 1 if there is some available point in the interval that is out of the region. In other case the result will be 0.
- For true_percentage, the result will be 1 if the series has already enough points in the interval inside the region. In other case the result will be 0.

## 5. Early Classification

The capability of obtaining an initial classification as soon as the available data give some cue about the possible class they belong to, is a very desirable property if the classifiers are to be used in a dynamic environment were data are generated and analyzed on line. Now, the setting is different to the variable length series learning problem of the previous paragraph: the classifier has already been obtained, maybe form series of different length, and the variations occur on the length of the series to be classified, from partial examples to full series.

Somewhat surprisingly, this early classification ability can be obtained without modifying the learning method, exploiting the same ideas that allow learning from series of different length. When a partial example is presented to the classifier some of its literals can be evaluated, because their intervals refer to areas that are already available in the example. Certainly, some literals cannot be evaluated because the intervals that they refer to are still not available for the example: its value is still unknown. Given that the classifier consists of a linear combination of literals, the simple omission of literals with unknown value allows to obtain a classification from the available data. The classification given to a partial example will be the linear combination of those literals that have known results.

A very simple approach to identify literals with unknown values would be to abstain whenever there are point values that are still unknown for the example. Nevertheless, if the values of some points in the interval are known, it is sometimes possible to know the result of the literal. This is done in a similar way than when evaluating literals for variable length series.

## 6. Experimental Validation

The characteristics of the data sets used to evaluate the behavior of the learning systems are summarized in Table 2. If a training test partition it specified, the results are obtained by averaging 5 experiments. In other case, 10 fold stratified cross validation was used.

Table 3 shows the obtained results for different data sets with different combination of predicates. The setting named *Relative* uses the predicates increases, decreases and stays. The setting *Interval* also includes these literals, and true_percentage. It does not include always and sometimes, because they can be considered as special cases of true_percentage.

Normally, 100 literals were used. For some data sets this value is more than enough, and it shows clearly that the method does not overfit. Nevertheless, the results of some data set can be improved using more

Table 2. Characteristics of the data sets.

| | Classes | Variables | Examples | Training/ Test | Length | | |
|---|---|---|---|---|---|---|---|
| | | | | | Min. | Average | Max. |
| CBF | 3 | 1 | 798 | 10 fold CV | 128 | 128.00 | 128 |
| CBF-Var | 3 | 1 | 798 | 10 fold CV | 64 | 95.75 | 128 |
| Control | 6 | 1 | 600 | 10 fold CV | 60 | 60.00 | 60 |
| Control-Var | 6 | 1 | 600 | 10 fold CV | 30 | 44.86 | 60 |
| Trace | 16 | 4 | 1600 | 800/800 | 67 | 82.49 | 99 |
| Auslan | 10 | 8 | 200 | 10 fold CV | 33 | 49.34 | 102 |

Table 3. Results (error rates) for the different data sets using the different literals.

| Data set | Literals | Point | Relative | Always/ sometimes | True percentage | Interval |
|---|---|---|---|---|---|---|
| CBF | 100 | 3.51 | 2.28 | 1.38 | 0.63 | 0.50 |
| CBF-Var | 100 | 3.49 | 1.61 | 2.75 | 1.25 | 1.13 |
| Control | 100 | 4.00 | 1.17 | 1.00 | 1.33 | 0.83 |
| Control-Var | 100 | 13.00 | 7.00 | 5.00 | 4.83 | 5.00 |
| Trace | 100 | 72.00 | 3.00 | 3.83 | 8.70 | 0.78 |
| Trace | 200 | 70.98 | 0.03 | 2.65 | 6.78 | 0.20 |
| Gloves | 100 | 8.00 | 5.50 | 5.50 | 3.00 | 4.50 |
| Gloves | 200 | 7.50 | 6.50 | 3.50 | 2.50 | 2.50 |
| Gloves | 300 | 5.00 | 4.00 | 4.50 | 2.50 | 1.50 |

Table 4. Early Classification. Results (error rates) for the different data sets using the different literals.

| Data set | Percentage | Point | Relative | Always/ sometimes | True percentage | Interval |
|---|---|---|---|---|---|---|
| CBF | 60 | 4.51 | 4.65 | 2.63 | 1.39 | 0.88 |
| CBF | 80 | 4.26 | 2.70 | 2.01 | 0.50 | 0.75 |
| CBF-Var | 60 | 4.11 | 1.86 | 3.12 | 1.87 | 1.61 |
| CBF-Var | 80 | 3.49 | 1.87 | 2.75 | 1.25 | 1.00 |
| Control | 60 | 42.17 | 17.33 | 30.17 | 35.00 | 34.33 |
| Control | 80 | 11.67 | 3.50 | 1.67 | 1.17 | 1.00 |
| Control-Var | 60 | 22.50 | 20.67 | 20.33 | 19.00 | 17.17 |
| Control-Var | 80 | 13.00 | 7.17 | 5.00 | 5.33 | 5.00 |
| Trace | 60 | 83.33 | 32.00 | 55.48 | 47.73 | 37.10 |
| Trace | 80 | 70.95 | 0.45 | 5.45 | 7.68 | 0.63 |
| Gloves | 60 | 5.00 | 4.00 | 4.50 | 3.00 | 1.50 |
| Gloves | 80 | 5.00 | 4.00 | 4.50 | 2.50 | 1.50 |

literals. Table 3 also shows for these data sets the obtained results using more iterations.

An interesting issue of these results is that the error rate achieved using point based literals are always the worst; a detailed study of this topic for fixed length time series can be found in [Rodriguez *et al.* (2001)]. The obtained results for the setting named *Interval* are the better or very close to the better results. The comparison between relative and region based literal clearly indicates that its adequacy depends on the data set.

Table 4 shows some results for early classification. The considered lengths are expressed in terms of the percentage of the length of the longest series in the data set. The table shows early classification results for 60% and 80%. Again, the usefulness of interval literals is clearly confirmed.

The rest of this section contains a detailed discussion for each data set, including its description, the results for boosting point based and interval based literals and another known results for the data set.

## 6.1. *CBF (Cylinder, Bell and Funnel)*

This is an artificial problem, introduced in [Saito (1994)]. The learning task is to distinguish between three classes: cylinder ($c$), bell ($b$) or funnel ($f$). Examples are generated using the following functions:

$$c(t) = (6 + \eta) \cdot \chi_{[a,b]}(t) + \varepsilon(t),$$
$$b(t) = (6 + \eta) \cdot \chi_{[a,b]}(t) \cdot (t - a)/(b - a) + \varepsilon(t),$$
$$f(t) = (6 + \eta) \cdot \chi_{[a,b]}(t) \cdot (b - t)/(b - a) + \varepsilon(t)$$

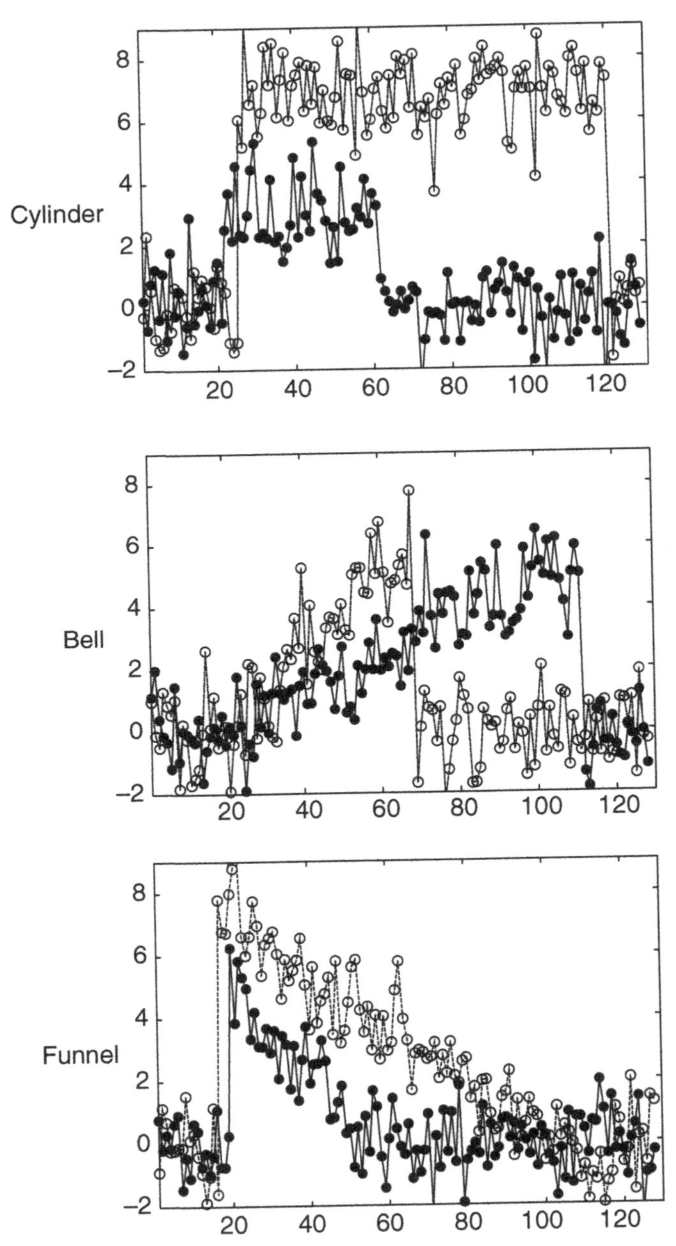


Fig. 3. Examples of the CBF data set. Two examples of the same class are shown in each graph.

where

$$\chi_{[a,b]} = \begin{cases} 0 & \text{if} \quad t < a \vee t > b, \\ 1 & \text{if} \quad a \leq t \leq b, \end{cases}$$

and $\eta$ and $\varepsilon(t)$ are obtained from a standard normal distribution $N(0, 1)$, $a$ is an integer obtained from a uniform distribution in [16, 32] and $b - a$ is another integer obtained from another uniform distribution in [32, 96]. The examples are generated evaluating those functions for $t = 1, 2, \ldots, 128$. Figure 3 shows some examples of this data set.

The obtained results are shown in Figure 4(A). Two kinds of graphs are used. The first one shows the error rate as a function of the number of literals. The second also shows the error rate, but in this case using early classification and the maximum number of considered literals. The error rate is shown as a function of the percentage of the length of the series. In each graph, two results are plotted, one using point based literals and another using interval based literals. The results are also shown in Table 5. For early classification, the results obtained with the 60% of the series length are very close to those obtained with the full series.

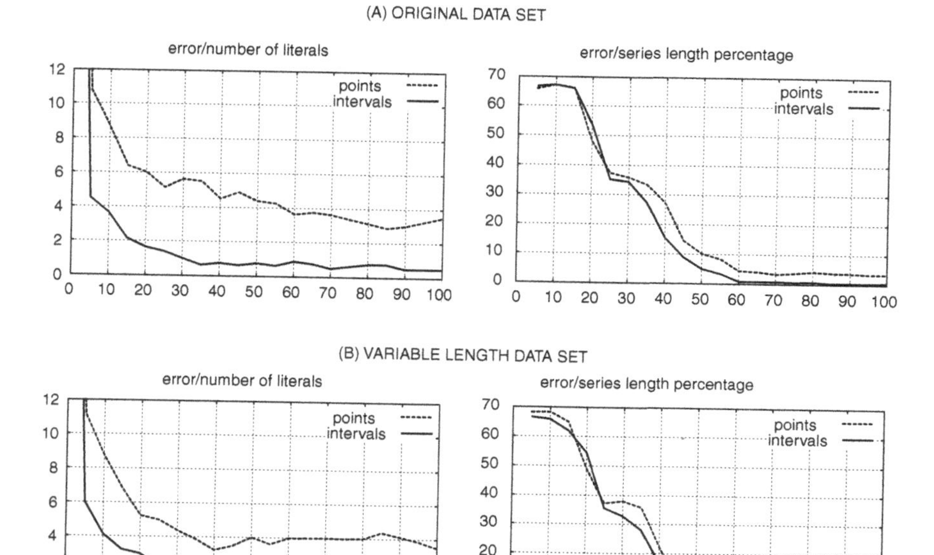


Fig. 4. Error graphs for the *CBF* data set.

Table 5. Results for the *CBF* data set.

| | 10 | 20 | 30 | 40 | 50 | 60 | 70 | 80 | 90 | 100 |
|---|---|---|---|---|---|---|---|---|---|---|
| *Error/Number of Literals* | | | | | | | | | | |
| Points | 8.67 | 6.02 | 5.64 | 4.51 | 4.39 | 3.64 | 3.64 | 3.15 | 3.02 | 3.51 |
| Intervals | 3.63 | 1.64 | 1.00 | 0.75 | 0.75 | 0.88 | 0.50 | 0.75 | 0.50 | 0.50 |
| *Error/Series Length Percentage* | | | | | | | | | | |
| Points | 67.17 | 48.15 | 35.85 | 27.10 | 10.13 | 4.51 | 3.39 | 4.26 | 3.77 | 3.51 |
| Intervals | 67.19 | 53.78 | 34.34 | 15.44 | 5.04 | 0.88 | 0.88 | 0.75 | 0.38 | 0.50 |

Table 6. Results for the *CBF-Var* data set.

| | 10 | 20 | 30 | 40 | 50 | 60 | 70 | 80 | 90 | 100 |
|---|---|---|---|---|---|---|---|---|---|---|
| *Error/Number of literals* | | | | | | | | | | |
| Points | 8.75 | 5.24 | 4.37 | 3.25 | 4.01 | 3.98 | 3.99 | 4.00 | 4.12 | 3.49 |
| Intervals | 4.11 | 2.99 | 2.13 | 1.75 | 1.50 | 1.25 | 1.13 | 1.25 | 1.13 | 1.13 |
| *Error/Series length percentage* | | | | | | | | | | |
| Points | 68.30 | 49.02 | 37.99 | 22.22 | 5.62 | 4.11 | 3.37 | 3.49 | 3.49 | 3.49 |
| Intervals | 65.84 | 54.38 | 32.83 | 16.67 | 3.25 | 1.61 | 1.00 | 1.00 | 1.13 | 1.13 |

The error reported in [Kadous (1999)] is 1.9, using event extraction, event clustering and decision trees. Our results using interval based literals are better than this value using 20 or more literals. Moreover, our method is able to obtain a better result than 1.9, 0.88, using early classification with 100 literals when only 60% of the series length is available.

**Variable Length Version.** All the examples of the *CBF* data set have the same length. In order to check the method for variable length series, the data set was altered. For each example, a random number of values were deleted from its end. The maximum number of values eliminated is 64, the half of the original length. The obtained results for this new data set, named *CBF-Var*, are shown in Figure 4(B) and in Table 6. When using a variable length data set and early classification, the series length percentage in graphs and tables are relative to the length of the *longest* series in the data set. For instance, in this data set 25% corresponds to the first 32 points of each series, because the longest series have 128 points.

An interesting issue is that the obtained results with this altered data set are still better than the results reported in [Kadous (1999)] using the original data set.

### 6.2. *Control Charts*

In this data set there are six different classes of control charts, synthetically generated by the process in [Alcock and Manolopoulos (1999)]. Each time series is of length $n = 60$, and it is defined by $y(t)$, with $1 \leq t \leq n$:

- Normal: $y(t) = m + rs$. Where $m = 30$, $s = 2$ and $r$ is a random number in [−3,3].
- Cyclic: $y(t) = m + rs + a\sin(2\pi t/T)$. $a$ and $T$ are in [10,15].
- Increasing: $y(t) = m + rs + gt$. $g$ is in [0.2,0.5].
- Decreasing: $y(t) = m + rs - gt$.
- Upward: $y(t) = m + rs + kx$. $x$ is in [7.5,20] and $k = 0$ before time $t_3$ and 1 after this time. $t_3$ is in $[n/3, 2n/3]$.
- Downward: $y(t) = m + rs - kx$.

Figure 5 shows two examples of each class. The data used was obtained from the UCI KDD Archive [Hettich and Bay (1999)]. The results are shown in Figure 6(A) and in Table 7. The results are clearly better for interval based literals, with the exception of early classification with very short fragments. This is possibly due to the fact that most of the literals refer to intervals that are after these initial fragments. In any case, early classification is not useful with so short fragments, because the error rates are too large.

**Variable Length Version.** The *Control* data set was also altered to a variable length variant, *Control-Var.* In this case the resulting series have lengths between 30 and 60. The results are shown in Figure 6(B) and in Table 8.

### 6.3. *Trace*

This dataset is introduced in [Roverso (2000)]. It is proposed as a benchmark for classification systems of temporal patterns in the process industry. This data set was generated artificially. There are four variables, and each variable has two behaviors, as shown in Figure 7. The combination of the behaviors of the variables produces 16 different classes. For this data set it is specified a training/test partition.

The length of the series is between 268 and 394. The running times of the algorithms depend on the length of the series, so the data set was preprocessed, averaging the values in groups of four. The lengths of the examples in the resulting data set have lengths between 67 and 99.

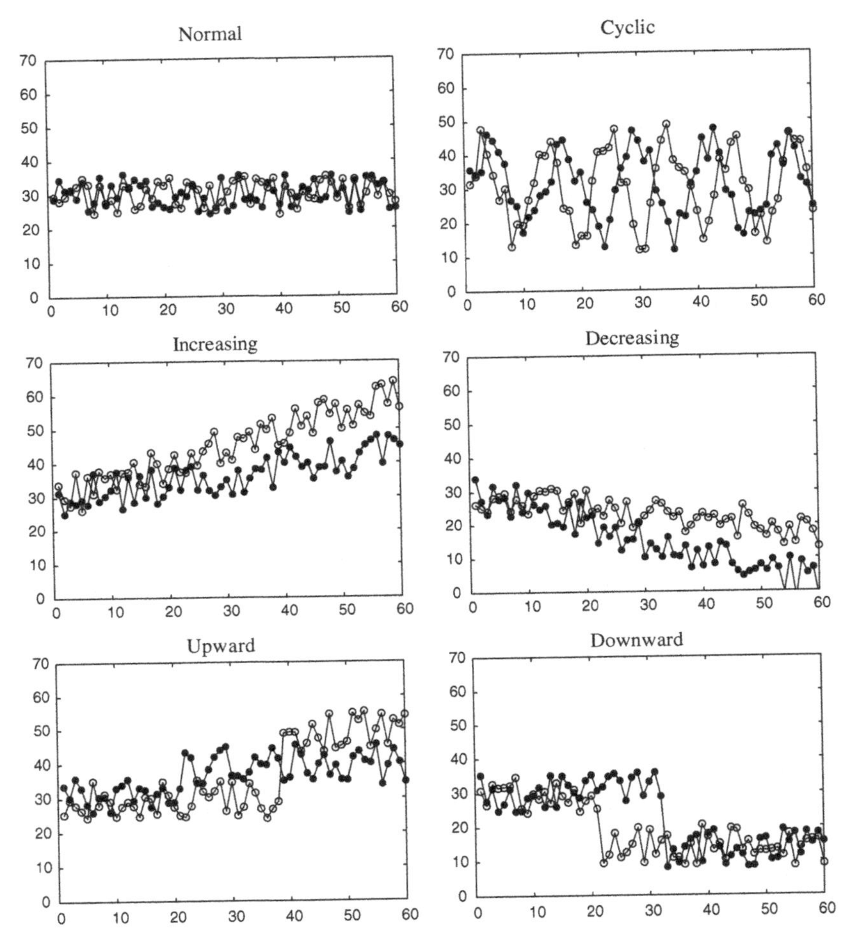


Fig. 5. Examples of the *Control* data set. Two examples of the same class are shown in each graph.

The results are shown in Figure 8 and in Table 9. For this data set, boosting point based literals is not able to learn anything useful, the error is always greater than 70%. Nevertheless, using boosting of interval based literals and more than 80 literals, the error is less than 1%. For early classification, the error is under 1% with the 80% of the series length. The result reported in [Roverso (2000)], using recurrent neural networks and wavelets is an error of 1.4%, but 4.5% of the examples are not assigned to any class.

(A) ORIGINAL DATASET

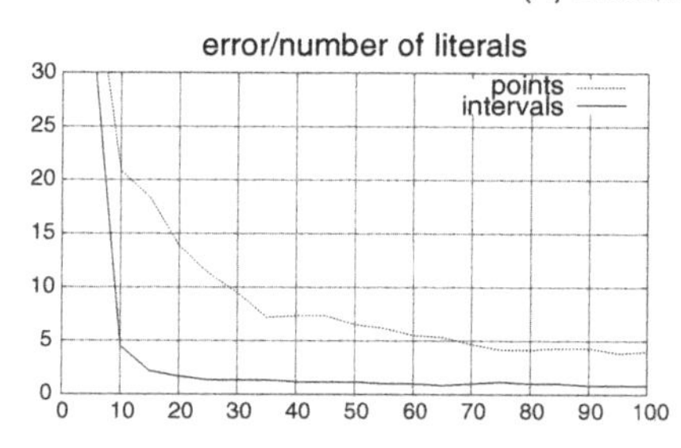


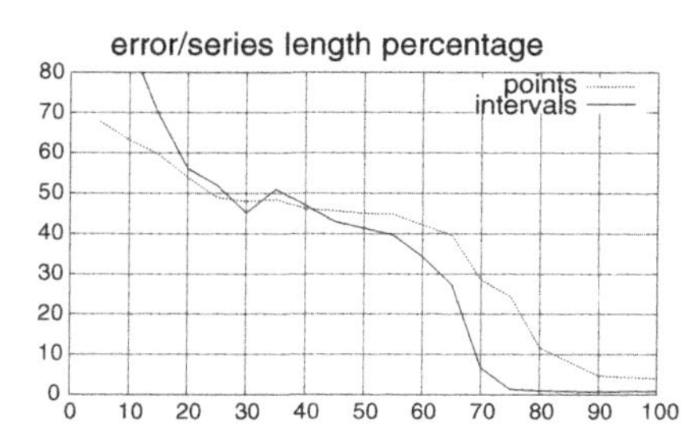


(B) VARIABLE LENGTH DATASET

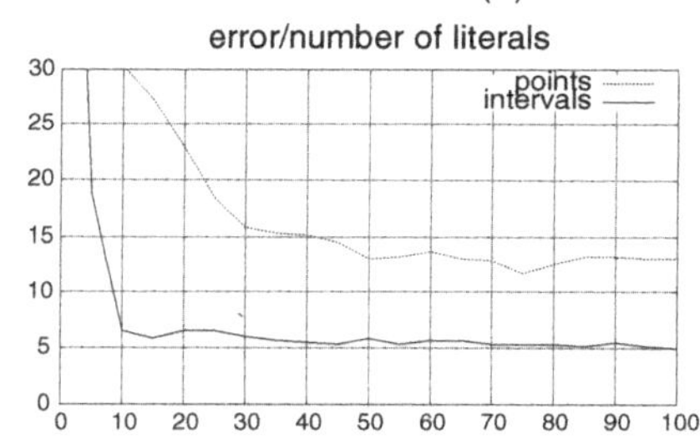


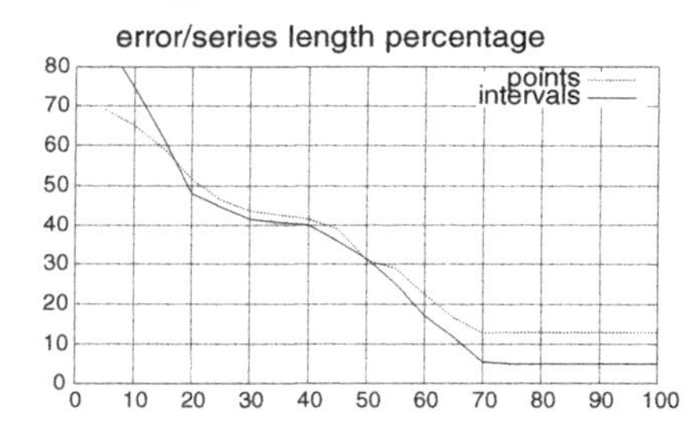


Fig. 6. Error graphs for the *Control* data set.

Table 7. Results for the *Control* data set.

| | 10 | 20 | 30 | 40 | 50 | 60 | 70 | 80 | 90 | 100 |
|---|---|---|---|---|---|---|---|---|---|---|
| *Error/Number of literals* | | | | | | | | | | |
| Points | 20.83 | 13.83 | 9.50 | 7.33 | 6.50 | 5.50 | 4.67 | 4.17 | 4.33 | 4.00 |
| Intervals | 4.50 | 1.67 | 1.33 | 1.17 | 1.17 | 1.00 | 1.00 | 1.00 | 0.83 | 0.83 |
| *Error/Series length percentage* | | | | | | | | | | |
| Points | 63.17 | 54.00 | 48.00 | 46.17 | 45.00 | 42.17 | 28.50 | 11.67 | 4.67 | 4.00 |
| Intervals | 87.00 | 56.17 | 45.17 | 47.00 | 41.33 | 34.33 | 6.67 | 1.00 | 0.67 | 0.83 |

Table 8. Results for the *Control-Var* data set.

| | 10 | 20 | 30 | 40 | 50 | 60 | 70 | 80 | 90 | 100 |
|---|---|---|---|---|---|---|---|---|---|---|
| *Error/Number of literals* | | | | | | | | | | |
| Points | 30.33 | 22.83 | 15.83 | 15.17 | 13.00 | 13.67 | 12.83 | 12.50 | 13.17 | 13.00 |
| Intervals | 6.50 | 6.50 | 6.00 | 5.50 | 5.83 | 5.67 | 5.33 | 5.33 | 5.50 | 5.00 |
| *Error/Series length percentage* | | | | | | | | | | |
| Points | 65.00 | 51.67 | 43.67 | 41.50 | 31.00 | 22.50 | 12.83 | 13.00 | 13.00 | 13.00 |
| Intervals | 74.67 | 48.17 | 41.50 | 40.00 | 31.33 | 17.17 | 5.50 | 5.00 | 5.00 | 5.00 |

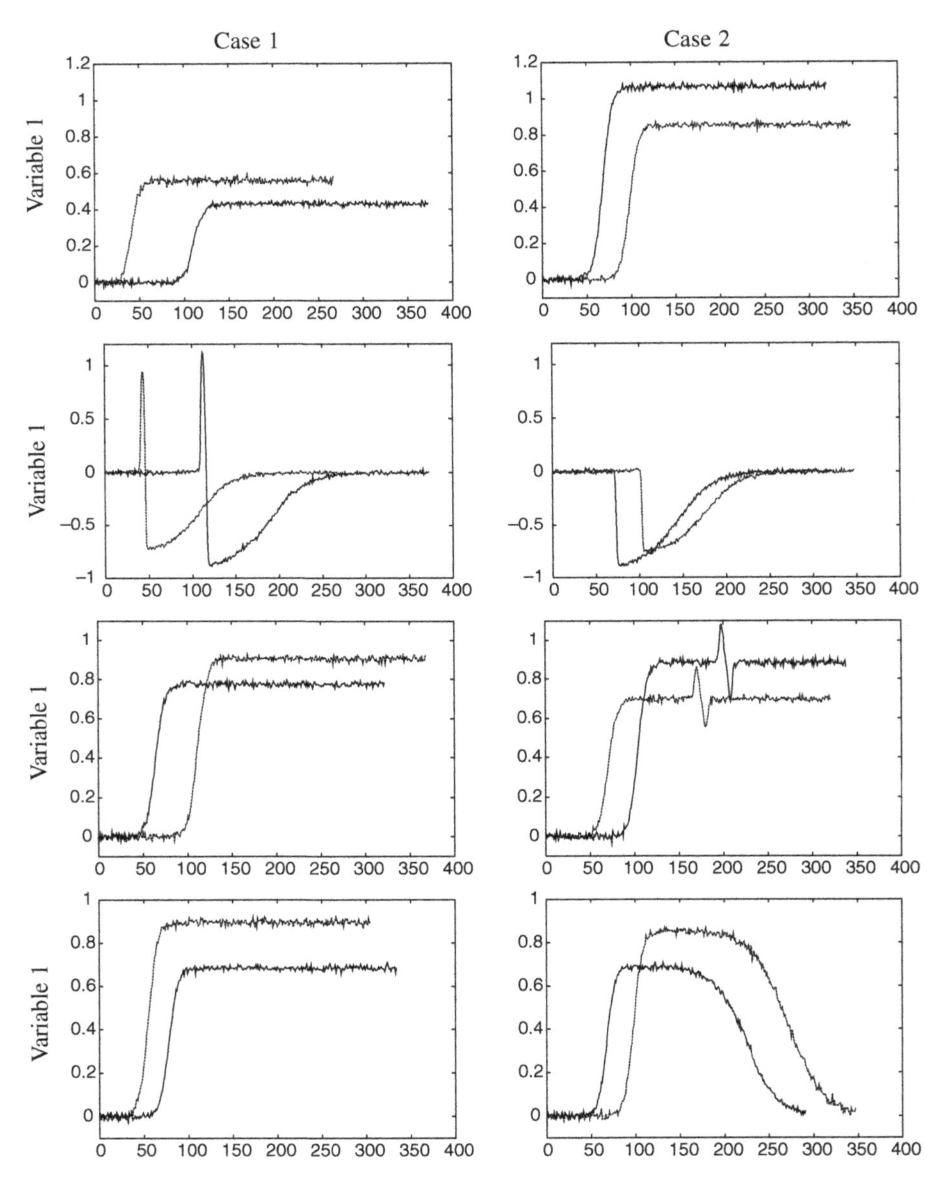


Fig. 7. Trace data set. Each example is composed by four variables, and each variable has two possible behaviors. In the graphs, two examples of each behavior are shown.

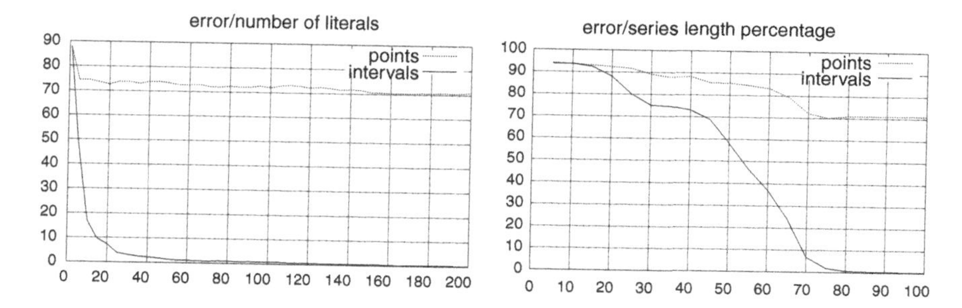


Fig. 8. Error graphs for the *Trace* data set.

Table 9. Results for the *Trace* data set.

| | Error/Number of literals | | | | | | | | | |
|---|---|---|---|---|---|---|---|---|---|---|
| Literals | 20 | 40 | 60 | 80 | 100 | 120 | 140 | 160 | 180 | 200 |
| Points | 72.63 | 73.83 | 72.78 | 72.40 | 72.00 | 72.25 | 71.70 | 70.45 | 70.55 | 70.98 |
| Intervals | 7.30 | 2.20 | 1.05 | 0.98 | 0.78 | 0.53 | 0.43 | 0.28 | 0.25 | 0.20 |
| | Error/Series length percentage | | | | | | | | | |
| Percentage | 10 | 20 | 30 | 40 | 50 | 60 | 70 | 80 | 90 | 100 |
| Points | 93.63 | 92.48 | 89.20 | 88.28 | 85.48 | 83.32 | 72.13 | 70.95 | 70.98 | 70.98 |
| Intervals | 93.75 | 88.15 | 75.05 | 73.05 | 58.05 | 37.10 | 6.90 | 0.63 | 0.28 | 0.20 |

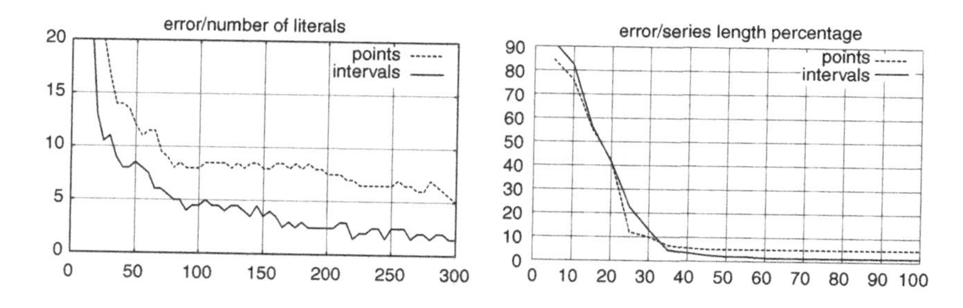


Fig. 9. Error graphs for the *Auslan* data set.

## 6.4. *Auslan*

Auslan is the Australian sign language, the language of the Australian deaf community. Instances of the signs were collected using an instrumented glove [Kadous (1999)]. Each example is composed by 8 series: $x$, $y$ and $z$ position, wrist roll, thumb, fore, middle and ring finger bend. There are

Table 10. Results for the *Auslan* data set.

| | Error/Number of literals | | | | | | | | | |
|---|---|---|---|---|---|---|---|---|---|---|
| Literals | 30 | 60 | 90 | 120 | 150 | 180 | 210 | 240 | 270 | 300 |
| Points | 16.50 | 11.50 | 8.00 | 8.50 | 8.00 | 8.00 | 7.50 | 6.50 | 6.00 | 5.00 |
| Intervals | 11.00 | 7.50 | 4.00 | 4.00 | 3.50 | 3.00 | 3.00 | 2.50 | 2.00 | 1.50 |
| | Error/Series length percentage | | | | | | | | | |
| Percentage | 10 | 20 | 30 | 40 | 50 | 60 | 70 | 80 | 90 | 100 |
| Points | 76.00 | 42.00 | 10.00 | 5.50 | 5.00 | 5.00 | 5.00 | 5.00 | 5.00 | 5.00 |
| Intervals | 82.50 | 41.50 | 13.00 | 3.50 | 2.00 | 1.50 | 1.50 | 1.50 | 1.50 | 1.50 |

10 classes and 20 examples of each class. The minimum length is 33 and the maximum is 102.

The results are shown in Figure 9 and in Table 10. The best result reported in [Kadous (1999)] is an error of 2.50, using event extraction, event clustering and Naïve Bayes Classifiers. Our result is 1.5 using 300 interval based literals.

With respect to the results for early classification, it is necessary to consider two facts. First, for this problem it is not practical to use early classification, because the examples are generated very quickly. The second fact is that the used percentage is for the longest series, and in this data set the lengths vary a lot. Hence, the obtained results for a length of 50%, include many examples that are already completed.

## 7. Conclusions

This chapter has presented a novel approach to the induction of multivariate time series classifiers. It summarizes a supervised learning method that works boosting very simple base classifiers. From only one literal, the base classifiers, boosting creates classifiers consisting of a linear combination of literals.

The learning method is highly domain independent, because it only makes use of very general techniques, like boosting, and only employs very generic descriptions of the time series, interval based literals. In this work we have resorted to two kind of interval predicates: relative and region based. Relative predicates consider the evolution of the values in the interval, while region based predicates consider the occurrence of the values of a variable in a region during an interval. These kind of predicates, specifically design for time series although essentially domain independent, allows the assembly of very accurate classifiers.

Experiments on different data sets show that in terms of error rate the proposed method is highly competitive with previous approaches. On several data sets, it achieves better than all previously reported results we are aware of. Moreover, although the strength of the method is based on boosting, the experimental results using point based predicates show that the incorporation of interval predicates can improve significantly the obtained classifiers, especially when using less iterations.

An important aspect of the learning method is that it can deal directly with variable length series. As we have shown, the simple mechanism of allowing that the evaluation of a literal may give as a result an abstention when there are not enough data to evaluate the literal, make possible to learn from series of different length. The symbolic nature of the base classifiers facilitates their capacity to abstain. It also requires the use of a boosting method able to work with abstentions in the base classifiers.

Another feature of the method is the ability to classify incomplete examples. This early classification is indispensable for some applications, when the time necessary to generate an example can be rather big, and where it is not an option to wait until the full example is available. It is important to notice that this early classification does not influence the learning process. Particularly, it is relevant the fact that in order to get early classifiers there are neither necessary additional classifiers nor more complex classifiers. This early classification capacity is another consequence of the symbolic nature of the classifier. Nevertheless, an open issue is if it would be possible to obtain better results for early classification by modifying the learning process. For instance, literals with later intervals could be somewhat penalized.

An interesting advantage of the method is its simplicity. From a user point of view, the method has only one free parameter, the number of iterations. Moreover, the classifiers created with more iterations includes the previously obtained. Hence, it is possible (i) to select only an initial fragment of the final classifier and (ii) to continue adding literals to a previously obtained classifier. Although less important, from the programmer point of view the method is also rather simple. The implementation of boosting stumps is one of the easiest among classification methods.

The main focus of this work has been classification accuracy, at the expenses of classifier comprehensibility. There are methods that produce more comprehensible models than weighted literals, such as decision trees or rules, and literals over intervals can also be used with these models. A first attempt to learn rules of literals is described in [Rodríguez *et al.* (2000)], where we obtained less accurate but more comprehensible classifiers. Nevertheless, the use of these literals with other learning models requires more research.

Finally, a short remark about the suitability of the method for time series classification. Since the only properties of the series that are considered for their classification are the attributes tested by the interval based literals, the method will be adequate as long as the series may be discriminated according to what happens in intervals. Hence, although the experimental results are rather good, they cannot be generalized for arbitrary data sets. Nevertheless, this learning framework can be used with other kinds of literals, more suited for the problem at hand. For instance [Rodríguez and Alonso (2000)] uses literals based on distances between series.

## References


1. Alcock, R.J. and Manolopoulos, Y. (1999). Time-Series Similarity Queries Employing a Feature-Based Approach. In *Proceedings of the 7th Hellenic Conference on Informatics, Ioannina, Greece.*
2. Alonso González, C.J. and Rodríguez Diez, J.J. (1999). A Graphical Rule Language for Continuous Dynamic Systems. In *International Conference on Computational Intelligence for Modelling, Control and Automation (CIMCA'99), Vienna, Austria.*
3. Berndt, D. and Clifford, J. (1996). Finding Patterns in Time Series: A Dynamic Programming Approach. In Fayyad, U., Piatetsky-Shapiro, G., Smyth, P., and Uthurusamy, R., eds, *Advances in Knowledge Discovery and Data Mining*, pp. 229–248. AAAI Press/MIT Press.
4. Hettich, S. and Bay, S.D. (1999). The UCI KDD archive. http://kdd.ics.uci.edu/.
5. Kadous, M.W. (1999). Learning Comprehensible Descriptions of Multivariate Time Series. In Bratko, I. and Dzeroski, S., eds, *Proceedings of the 16th International Conference of Machine Learning (ICML-99).* Morgan Kaufmann.
6. Kubat, M., Koprinska, I., and Pfurtscheller, G. (1998). Learning to Classify Biomedical Signals. In Michalski, R., Bratko, I., and Kubat, M., eds, Machine Learning and Data Mining, pp. 409–428. John Wiley & Sons.
7. Rodríguez, J.J. and Alonso, C.J. (2000). Applying Boosting to Similarity Literals for Time Series Classification. In *Proceedings of the 1st International Workshop on Multiple Classifier Systems (MCS 2000).* Springer.
8. Rodríguez, J.J., Alonso, C.J., and Boström, H. (2000). Learning First Order Logic Time Series Classifiers: Rules and Boosting. In *Proceedings of the 4th European Conference on Principles of Data Mining and Knowledge Discovery (PKDD 2000).* Springer.

9. Rodríguez, J.J., Alonso, C.J., and Boström, H. (2001). Boosting Interval Based Literals. *Intelligent Data Analysis*, **5**(3), 245–262.
10. Roverso, D. (2000). Multivariate Temporal Classification by Windowed Wavelet Decomposition and Recurrent Neural Networks. In *3rd ANS International Topical Meeting on Nuclear Plant Instrumentation, Control and Human-Machine Interface.*
11. Saito, N. (1994). Local Feature Extraction and its Applications Using a Library of Bases. PhD thesis, Department of Mathematics, Yale University.
12. Schapire, R.E. (1999). A Brief Introduction to Boosting. In *16th International Joint Conference on Artificial Intelligence (IJCAI-99)*, pp. 1401–1406. Morgan Kaufmann.
13. Schapire, R.E. and Singer, Y. (1998). Improved Boosting Algorithms Using Confidence-Rated Predictions. In *11th Annual Conference on Computational Learning Theory* (*COLT 1998*), pp. 80–91. ACM.

# CHAPTER 8

# MEDIAN STRINGS: A REVIEW

Xiaoyi Jiang
*Department of Electrical Engineering and Computer Science*
*Technical University of Berlin*
*Franklinstrasse 28/29, D-10587 Berlin, Germany*
E-mail: jiang@iam.unibe.ch

Horst Bunke
*Department of Computer Science, University of Bern*
*Neubrückstrasse 10, CH-3012 Bern, Switzerland*
Email: bunke@iam.unibe.ch

Janos Csirik
*Department of Computer Science, Attila Jozsef University*
*Arpad tér 2, H-6720 Szeged, Hungary*

Time series can be effectively represented by strings. The median concept is useful in various contexts. In this chapter its adaptation to the domain of strings is discussed. We introduce the notion of median string and provide related theoretical results. Then, we give a review of algorithmic procedures for efficiently computing median strings. Some experimental results will be reported to demonstrate the median concept and to compare some of the considered algorithms.

*Keywords*: String distance; set median string; generalized median string; online handwritten digits.

## 1. Introduction

Strings provide a simple and yet powerful representation scheme for sequential data. In particular time series can be effectively represented by strings. Numerous applications have been found in a broad range of fields including computer vision [2], speech recognition, and molecular biology [13,34].

A large number of operations and algorithms have been proposed to deal with strings [1,5,13,34,36]. Some of them are inherent to the special nature of strings such as the shortest common superstring and the longest common substring, while others are adapted from other domains.

In data mining, clustering and machine learning, a typical task is to represent a set of (similar) objects by means of a single prototype. Interesting applications of the median concept have been demonstrated in dealing with 2D shapes [16, 33], binary feature maps [23], 3D rotation [9], geometric features (points, lines, or 3D frames) [32], brain models [12], anatomical structures [37], and facial images [31]. In this paper we discuss the adaptation of the median concept to the domain of strings.

The median concept is useful in various contexts. It represents a fundamental quantity in statistics. In sensor fusion, multisensory measurements of some quantity are averaged to produce the best estimate. Averaging the results of several classifiers is used in multiple classifier systems in order to achieve more reliable classifications.

The outline of the chapter is as follows. We first formally introduce the median string problem in Section 2 and provide some related theoretical results in Section 3. Sections 4 and 5 are devoted to algorithmic procedures for efficiently computing set median and generalized median strings. In Section 6 we report experimental results to demonstrate the median concept and to compare some of the considered algorithms. Finally, some discussions conclude this paper.

## 2. Median String Problem

Assuming an alphabet $\Sigma$ of symbols, a string $x$ is simply a sequence of symbols from $\Sigma$, i.e. $x = x_1x_2\cdots x_n$ where $x_i \in \Sigma$ for $i = 1,\ldots,n$. Given the space $U$ of all strings over $\Sigma$, we need a distance function $d(p,q)$ to measure the dissimilarity between two strings $p,q \in U$. Let $S$ be a set of $N$ strings from $U$. The essential information of $S$ is captured by a string $\bar{p} \in U$ that minimizes the sum of distances of $\bar{p}$ to all strings from $S$, also called the consensus error $E_S(p)$:

$$\bar{p} = \arg\min_{p\in U} E_S(p), \quad \text{where} \quad E_S(p) = \sum_{q\in S} d(p,q).$$

The string $\bar{p}$ is called a *generalized median* of $S$. If the search is constrained to the given set $S$, the resultant string

$$\hat{p} = \arg\min_{p\in S} E_S(p)$$

is called a *set median* of $S$. For a given set $S$, neither the generalized median nor the set median is necessarily unique. This definition was introduced by Kohonen [20]. Note that different terminology has been used in the literature. In [13] the set median string and the generalized median string are termed *center string* and *Steiner string*, respectively. In [24] the generalized median was called consensus sequence.

Different possibility is mentioned by Kohonen [20] too. This is the parallel of mean from elementary statistics. Here we would like to search for $p'$ that minimizes

$$\sum_{q \in S} d^2(p', q).$$

Martinez-Hinarejos *et al.* [27] returned to this definition and investigated the possibility of using mean instead of median.

Note that in a broad sense, the median string concept is related to the center of gravity of a collection of masses. While the latter represents the point where all the weight of the masses can be considered to be concentrated, the median string corresponds to a single representation of strings based on a string distance.

Several string distance functions have been proposed in the literature. The most popular one is doubtlessly the Levenshtein edit distance. Let $A = a_1 a_2 \cdots a_n$ and $B = b_1 b_2 \cdots b_m$ be two words over $\Sigma$. The Levenshtein edit distance $d(A, B)$ is defined in terms of elementary edit operations which are required to transform $A$ into $B$. Usually, three different types of edit operations are considered, namely (1) substitution of a symbol $a \in A$ by a symbol $b \in B, a \neq b$, (2) insertion of a symbol $a \in \Sigma$ in $B$, and (3) deletion of a symbol $a \in A$. Symbolically, we write $a \to b$ for a substitution, $\varepsilon \to a$ for an insertion, and $a \to \varepsilon$ for a deletion. To model the fact that some distortions may be more likely than others, costs of edit operations, $c(a \to b), c(\varepsilon \to a)$, and $c(a \to \varepsilon)$, are introduced. Let $s = l_1 l_2 \cdots l_k$ be a sequence of edit operations transforming $A$ into $B$. We define the cost of this sequence by $c(s) = \sum_{i=1}^{k} c(l_i)$. Given two strings $A$ and $B$, the Levenshtein edit distance is given by

$$d(A, B) = \min\{c(s) \mid s \text{ is a sequence of edit operations transforming } A \text{ into } B\}.$$

To illustrate the Levenshtein edit distance, let us consider two words $A = median$ and $B = mean$ built on the English alphabet. Examples of sequences of edit operations transforming $A$ into $B$ are:

- $s_1 = d \to a, \quad i \to n, \quad a \to \varepsilon, \quad n \to \varepsilon,$
- $s_2 = d \to a, \quad i \to \varepsilon, \quad a \to \varepsilon,$
- $s_3 = d \to \varepsilon, \quad i \to \varepsilon.$

Under the edit cost $c(a \to \varepsilon) = c(\varepsilon \to a) = c(a \to b) = 1$, $a \neq b$, $s_3$ represents the optimal sequence with the minimum total cost 2 for transforming *median* into *mean* among all possible transformations. Therefore, we observe $d(median, mean) = 2$.

In [38] an algorithm is proposed to compute the Levenshtein edit distance by means of dynamic programming. Other algorithms are discussed in [13,34,36]. Further string distance functions are known from the literature, for instance, normalized edit distance [28], maximum posterior probability distance [20], feature distance [20], and others [8]. The Levenshtein edit distance is by far the most popular one. Actually, some of the algorithms we discuss later are tightly coupled to this particular distance function.

## 3. Theoretical Results

In this section we summarize some theoretical results related to median strings. The generalized median is a more general concept and usually a better representation of the given strings than the set median. From a practical point of view, the set median can be regarded an approximate solution of the generalized median. As such it may serve as the start for an iterative refinement process to find more accurate approximations. Interestingly, we have the following result (see [13] for a proof):

**Theorem 1.** *Assume that the string distance function satisfies the triangle inequality. Then* $E_S(\hat{p})/E_S(\bar{p}) \leq 2 - 2/|S|$.

That is, the set median has a consensus error relative to $S$ that is at most $2 - 2/|S|$ times the consensus error of the generalized median string.

Independent of the distance function we can always find the set median of $N$ strings by means of $\frac{1}{2}N(N-1)$ pairwise distance computations. This computational burden can be further reduced by making use of special properties of the distance function (e.g. metric) or resorting to approximate procedures. Section 4 will present examples of these approaches.

Compared to set median strings, the computation of generalized median strings represents a much more demanding task. This is due to the huge search space which is substantially larger than that for determining the set median string. This intuitive understanding of the computational complexity is supported by the following theoretical results. Under the two conditions:

- every edit operation has cost one, i.e., $c(a \to b) = c(\varepsilon \to a) = c(a \to \varepsilon) = 1$,
- the alphabet is not of fixed size.

It is proved in [15] that computing the generalized median string is NP-hard. Sim and Park [35] proved that the problem is NP-hard for finite alphabet and for a metric distance matrix. Another result comes from computational biology. The optimal evolutionary tree problem there turns out to be equivalent to the problem of computing generalized median strings if the tree structure is a star (a tree with $n+1$ nodes, $n$ of them being leaves). In [39] it is proved that in this particular case the optimal evolutionary tree problem is NP-hard. The distance function used is problem dependent and does not even satisfy the triangle inequality. All these theoretical results indicate the inherent difficulty in finding generalized median strings. Not surprisingly, the algorithms we will discuss in Section 5 are either exponential or approximate.

## 4. Fast Computation of Set Median Strings

The naive computation of set median requires $O(N^2)$ distance computations. Considering the relatively high computational cost of each individual string distance, this straightforward approach may not be appropriate, especially in the case of a large number of strings. The problem of fast set median search can be tackled by making use of properties of metric distance functions or developing approximate algorithms. Several solutions [19,30] have been suggested for fast set median search in arbitrary spaces. They apply to the domain of strings as well.

### 4.1. *Exact Set Median Search in Metric Spaces*

In many applications the underlying string distance function is a metric which satisfies:

(i) $d(p,q) \geq 0$ and $d(p,q) = 0$ if and only if $p = q$,
(ii) $d(p,q) = d(q,p)$,
(iii) $d(p,q) + d(q,r) \geq d(p,r)$.

A property of metrics is:

$$|d(p,r) - d(r,q)| \leq d(p,q), \quad \forall p,q,r \in S,$$

which can be utilized to reduce the number of string distance computations.

The approach proposed in [19] partitions the input set $S$ into subsets $S_u$ (used), $S_e$ (eliminated), and $S_a$ (alive). The set $S_a$ keeps track of those strings that have not been fully evaluated yet; initially $S_a = S$. A lower bound $g(p)$ is computed for each string $p$ in $S_a$, i.e., the consensus error of $p$ satisfies:

$$E_S(p) = \sum_{q \in S} d(p,q) \geq g(p).$$

Clearly, strings with small $g$ values are potentially better candidates for set median. For this reason the string $p$ with the smallest $g(p)$ value among all strings in $S_a$ is transferred from $S_a$ to $S_u$. Then, the consensus error $E_S(p)$ is computed and, if necessary, the current best median candidate $p^*$ is updated. Then, the lower bound $g$ is computed for all strings that are alive, and those whose $g$ is not smaller than $E_S(p^*)$ are moved from $S_a$ to $S_e$. They will not be considered as median candidates any longer. This process is repeated until $S_a$ becomes empty.

In each iteration, the consensus error for $p$ with the smallest $g$ value is computed by:

$$E_S(p) = \sum_{q \in S_u} d(p,q) + \sum_{q \in S_e \cup (S_a - \{p\})} d(p,q).$$

Using (1) the term $d(p,q)$ in the second summation is estimated by:

$$d(p,q) \geq |d(p,r) - d(r,q)|, \quad \forall r \in S_u.$$

Taking all strings of $S_u$ into account, we obtain the lower bound:

$$E_S(p) \geq \sum_{q \in S_u} d(p,q) + \sum_{q \in S_e \cup (S_a - \{p\})} \max_{r \in S_u} |d(p,r) - d(r,q)| = g(p).$$

The critical point here is to see that all the distances in this lower bound are concerned with $p$ and strings from $S_u$, and were therefore already computed before. When strings in $S_a$ are eliminated (moved to $S_u$), their consensus errors need not to be computed in future. This fact results in saving of distance computations. In addition to (2), two other lower bounds within the same algorithmic framework are given in [19]. They differ in the resulting ratio of the number of distance computations and the remaining overhead, with the lower bound (2) requiring the smallest amount of distance computations.

Ideally, the distance function is desired to be a metric, in order to match the human intuition of similarity. The triangle inequality excludes the case in which $d(p,r)$ and $d(r,q)$ are both small, but $d(p,q)$ is very large. In practice, however, there may exist distance functions which do not satisfy the triangle inequality. To judge the suitability of these distance functions, the work [6] suggests quasi-metrics with a relaxed triangle inequality. Instead of the strict triangle inequality, the relation:

$$d(p,r) + d(r,q) \geq \frac{d(p,q)}{1+\varepsilon}$$

is required now. Here $\varepsilon$ is a small nonnegative constant. As long as $\varepsilon$ is not very large, the relaxed triangle inequality still retains the human intuition of similarity. Note that the strict triangle inequality is a special case with $\varepsilon = 0$. The fast set median search procedure [19] sketched above easily extends to quasi-metrics. In this case the relationship (1) is replaced by:

$$d(p,q) \geq \max\left(\frac{d(p,r)}{1+\varepsilon} - d(q,r), \frac{d(q,r)}{1+\varepsilon} - d(p,r)\right), \quad \forall p,q,r \in S$$

which can be used in the same manner to establish a lower bound $g(p)$.

### 4.2. *Approximate Set Median Search in Arbitrary Spaces*

Another approach to fast set median search makes no assumption on the distance function and covers therefore non-metrics as well. The idea of this approximate algorithm is simple. Instead of computing the sum of distances of each string to all the other strings of $S$ to select the best one, only a subset of $S$ is used to obtain an estimation of the consensus error [30]. The algorithm first calculates such estimations and then calculates the exact consensus errors only for strings that have low estimations.

This approximate algorithm proceeds in two steps. First, a random subset $S_r$ of $N_r$ strings is selected from $S$. For each string $p$ of $S$, the consensus error $E_{S_r}(p)$ relative to $S_r$ is computed and serves as an estimation of the consensus error $E_S(p)$. In the second step $N_t$ strings with the lowest consensus error estimations are chosen. The exact consensus error $E_S(p)$ is computed for the $N_t$ strings and the string with the minimum $E_S(p)$ is regarded the (approximate) set median string of $S$.

## 5. Computation of Generalized Median Strings

While the set median problem is characterized by selecting one particular member out of a given set of strings, the computation of generalized median strings is inherently constructive. The theoretical results from Section 3 about computational complexity indicate the fundamental difficulties we are faced with. In the following we describe various algorithms for computing generalized median strings. Not surprisingly, they are either of exponential complexity or approximate.

### 5.1. *An Exact Algorithm and Its Variants*

An algorithm for the exact computation of generalized median strings under the Levenshtein distance is given in [21]. Let $\varepsilon$ be the empty string and $\Sigma' = \Sigma \cup \{\varepsilon\}$ the extended alphabet. We define:

$$\delta(r_1, r_2, \ldots, r_N) = \min_{v \in \Sigma'} [c(v \to r_1) + c(v \to r_2) + \cdots + c(v \to r_N)].$$

The operator $\delta$ can be interpreted as a voting function, as it determines the best value $v$ at a given stage of computation. Finding an optimal value of $v$ requires an exhaustive search over $\Sigma'$ in the most general case, but in practice the cost function is often simple such that a shortcut can be taken and the choice of the optimal $v$ is not costly.

Having defined $\delta$ this way, the generalized median string can be computed by means of dynamic programming in an $N$-dimensional array, similarly to string edit distance computation [38]. For the sake of notational simplicity, we only discuss the case $N = 3$. Assume the three input strings be $u_1 u_2 \cdots u_l$, $v_1 v_2 \cdots v_m$, and $w_1 w_2 \cdots w_n$. A three-dimensional distance table of dimension $l \times m \times n$ is constructed as follows:

**Initialization:** $d_{0,0,0} = 0$;
**Iteration:**

$$d_{i,j,k} = \min \left\{ \begin{array}{ll} d_{i-1,j-1,k-1} & + \delta(u_i, v_j, w_k) \\ d_{i-1,j-1,k} & + \delta(u_i, v_j, \varepsilon) \\ d_{i-1,j,k-1} & + \delta(u_i, \varepsilon, w_k) \\ d_{i-1,j,k} & + \delta(u_i, \varepsilon, \varepsilon) \\ d_{i,j-1,k-1} & + \delta(\varepsilon, v_j, w_k) \\ d_{i,j-1,k} & + \delta(\varepsilon, v_j, \varepsilon) \\ d_{i,j,k-1} & + \delta(\varepsilon, \varepsilon, w_k) \end{array} \right\} \left\{ \begin{array}{l} 0 \le i \le l \\ 0 \le j \le m \\ 0 \le k \le n \end{array} \right\}$$

**End:** if $(i = l) \wedge (j = m) \wedge (k = n)$

The computation requires $O(lmn)$ time and space. The path in the distance table that leads from $d_{0,0,0}$ to $d_{l,m,n}$ defines the generalized median string $\bar{p}$ with $d_{l,m,n}$ being the consensus error $E_S(\bar{p})$. Note that a generalization to arbitrary $N$ is straightforward. If the strings of $S$ are of length $O(n)$, both the time and space complexity amounts to $O(n^N)$ in this case.

Despite of its mathematical elegance the exact algorithm above is impractical because of the exponential complexity. There have been efforts to shorten the computation time using heuristics or domain-specific knowledge. Such an approach from [24] assumes that the string of $S$ be quite similar. Under reasonable constraints on the cost function ($c(a \to \varepsilon) = c(\varepsilon \to a) = 1$ and $c(a \to b)$ nonnegative), the generalized median string $\bar{p}$ satisfies $E_S(\bar{p}) \leq k$ with $k$ being a small number. In this case the optimal dynamic programming path must be close to the main diagonal in the distance table. Therefore only part of the $n$-dimensional table needs to be considered. Details of the restricted search can be found in [24]. Its asymptotic time complexity is $O(nk^N)$. While this remains exponential, $k$ is typically much smaller than $n$, resulting in a substantial speedup compared to the full search of the original algorithm [21].

We may also use any domain-specific knowledge to limit the search space. An example is the approach in the context of classifier combination for handwritten sentence recognition [25]. An ensemble of classifiers provide multiple classification results of a scanned text. Then, the consensus string is expected to yield the best overall recognition performance. The input strings from the individual classifiers are associated with additional information of position, i.e. the location of each individual word in a sequence of handwritten words. Obviously, it is very unlikely that a word at the beginning of a sequence corresponds to a word at the end of another sequence. More generally, only words at a similar position in the text image are meaningful candidates for being matched to each other. The work [25] makes use of this observation to exclude a large portion of the full $N$-dimensional search space from consideration.

## 5.2. *Approximate Algorithms*

Because of the NP-hardness of generalized median string computation, efforts have been undertaken to develop approximate approaches which provide suboptimal solutions in reasonable time. In this section we will discuss several algorithms of this class.

### 5.2.1. *Greedy Algorithms*

The following algorithm was proposed by Casacuberta and de Antoni [4]. Starting from an empty string, a greedy algorithm constructs an

$\bar{p}_0 = \varepsilon;$
**for** $l = 1;; l++$ {
  $a_1 = \arg\min_{a \in \Sigma} E_S(\bar{p}_{l-1}a);$
  $\bar{p}_l = \bar{p}_{l-1}a_l;$
  **if** (termination criterion fulfilled) **break**;
}
choose prefix of $\bar{p}_l$ for output;

Fig. 1. General framework of greedy algorithms.

approximate generalized median string $\bar{p}$ symbol by symbol. When we are going to generate the $l$-th symbol $a_l$ ($l \geq 1$), the substring $a_1 \cdots a_{l-1}$ ($\varepsilon$ for $l = 1$) has already been determined. Then, each symbol from $\Sigma$ is considered as a candidate for $a_l$. All the candidates are evaluated and the final decision of $a_l$ is made by selecting the best candidate. The process is continued until a termination criterion is fulfilled.

A general framework of greedy algorithms is given in Figure 1. There are several possible choices for the termination criterion and the prefix. The greedy algorithm proposed in [4] stops the iterative construction process when $E_S(\bar{p}_l) > E_S(\bar{p}_{l-1})$. Then, $\bar{p}_{l-1}$ is regarded the approximate generalized median string. Alternatively, the author of [22] suggests the termination criterion $l = \max_{p \in S} |p|$. The output is the prefix of $\bar{p}_l$ with the smallest consensus error relative to $S$. For both variants a suitable data structure [4] enables a time complexity of $O(n^2 N|\Sigma|)$ for the Levenshtein distance. The space complexity amounts to $O(nN|\Sigma|)$.

In the general framework in Figure 1 nothing is stated about how to select $a_l$ if there exist multiple symbols from $\Sigma$ with the same value of $E_S(\bar{p}_{l-1}a)$. Besides a simple random choice, the history of the selection process can be taken into account to make a more reliable decision [22].

### 5.2.2. *Genetic Search*

Genetic search techniques are motived by concepts from the theory of biological evolution. They are general-purpose optimization methods that have been successfully applied to a large number of complex tasks. In [17] two genetic algorithms are proposed to construct generalized median strings. The first approach is based on a straightforward representation of strings in terms of chromosomes by one-dimensional arrays of varying lengths consisting of symbols from $\Sigma$. The string "median", for instance, is coded as chromosome

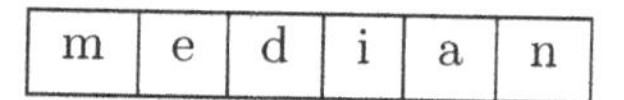

| m | e | d | i | a | n |
|---|---|---|---|---|---|

Given a set $S$ of $N$ input strings of length $O(n)$ each, the fitness function is simply defined by the consensus error of a chromosome string. The roulette wheel selection is used to create offspring. For crossover the single-point operator is applied. With a mutation probability either deletion, insertion or substitution is performed at each position of a chromosome. The input strings are used as part of the initial population. The rest of the initial population is filled by randomly selecting some input strings and applying the mutation operator on them. The population evolution process terminates if one of the following two criteria is fulfilled. The first criterion is that the maximal number of generations has been reached. Second, if the population becomes very homogeneous, the process terminates as well. The homogeneity is measured by means of the average and the variance of the fitness values of a population. If their product is smaller than a prespecified threshold, the population is considered enough homogeneous to stop the evolution. Let $P$ denote the population size and $p$ the replacement percentage. When using the Levenshtein edit distance, the time complexity of this genetic algorithm amounts to $O(Nn_2pP)$ per generation.

The genetic algorithm above requires quite a large number of string distance computations due to the fitness function evaluation in each iteration. For better computational efficiency a second genetic algorithm is designed using a more elaborate chromosome coding scheme of strings; see [17] for details. The time complexity becomes $O(NnmpP)$, $m << n$, implying a substantial speedup.

### 5.2.3. *Perturbation-Based Iterative Refinement*

The set median represents an approximation of the generalized median string. The greedy algorithms and the genetic search techniques also give approximate solutions. An approximate solution $\bar{p}$ can be further improved by an iterative process of systematic perturbations. This idea was first suggested in [20]. But no algorithmic details are specified there. A concrete algorithm for realizing systematic perturbations is given in [26]. For each position $i$, the following operations are performed:

(i) Build perturbations

- Substitution: Replace the $i$-th symbol of $\bar{p}$ by each symbol of $\Sigma$ in turn and choose the resulting string $x$ with the smallest consensus error relative to $S$.
- Insertion: Insert each symbol of $\Sigma$ in turn at the $i$-th position of $\bar{p}$ and choose the resulting string $y$ with the smallest consensus error relative to $S$.
- Deletion: Delete the $i$-th symbol of $\bar{p}$ to generate $z$.

(ii) Replace $\bar{p}$ by the one from $\{\bar{p}, x, y, z\}$ with the smallest consensus error relative to $S$.

For the Levenshtein distance one global iteration that handles all positions of the initial $\bar{p}$ needs $O(n^3 N|\Sigma|)$ time. The process is repeated until there is more improvement possible.

### 5.3. *Dynamic Computation of Generalized Median Strings*

In a dynamic context we are faced with the situation of a steady arrival of new data items, represented by strings. At each point of time, $t$, the set $S^t$ of existing strings is incremented by a new string, resulting in $S^{t+1}$, and its generalized median string is to be computed. Doubtlessly, a trivial solution consists in applying any of the approaches discussed above to $S^{t+1}$. By doing this, however, we compute the generalized median string of $S^{t+1}$ from scratch without utilizing any knowledge about $S^t$, in particular its generalized median string. All algorithms for computing generalized median strings are of such a static nature and thus not optimal in a dynamic context. The work [18] proposes a genuinely dynamic approach, in which the update scheme only considers the generalized median string of $S^t$ together with the new data item, but not the individual members of $S^t$.

The inspiration for the algorithm comes from a fundamental fact in real space. Under the distance function $d(p_i, p_j) = (p_i - p_j) \cdot (p_i - p_j)$, i.e. the squared Euclidean distance of $p_i$ and $p_j$, the generalized median of a given set $S^t = \{p_1, p_2, \ldots, p_t\}$ of $t$ points is the well-known mean:

$$\bar{p}^t = \frac{1}{t} \cdot \sum_{i=1}^{t} p_i.$$

When an additional point $p_{t+1}$ is added to $S^t$, the resultant new set $S^{t+1} = S^t \cup \{p_{t+1}\}$ has the generalized median

$$\bar{p}^{t+1} = \frac{1}{t+1} \cdot \sum_{i=1}^{t+1} p_i = \frac{t}{t+1} \cdot \bar{p}^t + \frac{1}{t+1} \cdot p_{t+1}$$

which is the so-called weighted mean of $\bar{p}^t$ and $p_{t+1}$ satisfying

$$d(\bar{p}^{t+1}, \bar{p}^t) = \frac{1}{t+1} \cdot d(\bar{p}^t, p_{t+1}),$$
$$d(\bar{p}^{t+1}, \bar{P}_{t+1}) = \frac{t}{t+1} \cdot d(\bar{p}^t, p_{t+1}).$$

Geometrically, $\bar{p}^{t+1}$ is a point on the straight line segment connecting $\bar{p}^t$ and $p_{t+1}$ that has distances $1/(t+1) \cdot d(\bar{p}^t, p_{t+1})$ and $1/(t+1) \cdot d(\bar{p}^t, p_{t+1})$ to $\bar{p}^t$ and $p_{t+1}$, respectively.

On a heuristic basis the special case in real space can be extended to the domain of strings. Given a set $S^t = \{p_1, p_2, \ldots, p_t\}$ of $t$ strings and its generalized median $\bar{p}^t$, the generalized median of a new set $S^{t+1} = S^t \cup \{p_{t+1}\}$ is estimated by a weighted mean of $\bar{p}^t$ and $p_{t+1}$, i.e. by a string $\bar{p}^{t+1}$ such that

$$d(\bar{p}^{t+1}, \bar{p}^t) = \alpha \cdot d(\bar{p}^t, p_{t+1}),$$
$$d(\bar{p}^{t+1}, p_{t+1}) = (1 - \alpha) \cdot d(\bar{p}^t, p_{t+1})$$

where $\alpha \in [0, 1]$. In real space $\alpha$ takes the value $1/(t+1)$. For strings, however, we have no possibility to specify $\alpha$ in advance. Therefore, we resort to a search procedure. Remember that our goal is to find $\bar{p}^{t+1}$ that minimizes the consensus error relative to $S^{t+1}$. To determine the optimal $\alpha$ value a series of $\alpha$ values $0, 1/k, \ldots, (k-1)/k$ is probed and the $\alpha$ value that results in the smallest consensus error is chosen.

The dynamic algorithm uses the method described in [3] for computing the weighted mean of two strings. It is an extension of the Levenshtein distance computation [38].

## 6. Experimental Evaluation

In this section we report some experimental results to demonstrate the median string concept and to compare some of the computational procedures described above. The used data are online handwritten digits from a subset[1] of the UNIPEN database [14]. An online handwritten digit is a time

[1] Available at ftp: //ftp.ics.uci.edu/pub/machine-learning-databases/pendigits/.

sequence of 2D points recorded from a person's writing on a special tablet. Each digit is originally given as a sequence $d = (x_1, y_1, t_1), \ldots, (x_k, y_k, t_k)$ of points in the $xy$-plane, where $t_k$ is the time of recording the $k$th point during writing a digit. In order to transform such a sequence of points into a string, we first resample the given data points such that the distance between any consecutive pair of points has a constant value $\Delta$. That is, $d$ is transformed into sequence $d' = (\bar{x}_1, \bar{y}_1), \ldots, (\bar{x}_l, \bar{y}_l)$ where $|(\bar{x}_{i+1}, \bar{y}_{i+1}) - (\bar{x}_i, \bar{y}_i)| = \Delta$ for $i = 1, \ldots, l-1$. Then, a string $s = a_1 a_2 \cdots a_{l-1}$ is generated from sequence $d'$ where $a_i$ is the vector pointing from $(\bar{x}_i, \bar{y}_i)$ to $(\bar{x}_{i+1}, \bar{y}_{i+1})$. In our experiments we fixed $\Delta = 7$ so that the digit strings have between 40 and 100 points.

The costs of the edit operations are defined as follows: $c(a \to \varepsilon) = c(\varepsilon \to a) = |a| = \Delta$, $c(a \to b) = |a - b|$. Notice that the minimum cost of a substitution is equal to zero (if and only if $a = b$), while the maximum cost is $2\Delta$. The latter case occurs if $a$ and $b$ are parallel and have opposite direction.

We conducted experiments using 10 samples of digit 1, 2 and 3 each, and 98 samples of digit 6, see Figure 2 for an illustration (for space reasons only ten samples of digit 6 are shown there). The results of four algorithms are shown in Figure 3: the genetic algorithm [17] (GA, Section 5.2.2), the dynamic algorithm [18] (Section 5.3), the greedy algorithm [4] (Section 5.2.1), and the set median. For comparison purpose the consensus error $E_S$ and the computation time $t$ in seconds on a SUN Ultra60 workstation are also listed in Figure 3. Note that the consensus errors of digit 6 are substantially larger than those of the other digits because of the definition of consensus error as the sum, but not the average, of the distances to all input samples.

The best results are achieved by GA, followed by the dynamic approach. Except for digit 1, the greedy algorithm reveals some weakness. Looking at the median for digits 2, 3 and 6 it seems that the iterative process terminates too early, resulting in a string (digit) much shorter than it should be. The reason lies in the simple termination criterion defined in [4]. It works well for the (short) words used there, but obviously encounters difficulties in dealing with longer strings occurring in our study. At first glance, the dynamic approach needs more computation time than the greedy algorithm. But one has to take into account that the recorded time is the total time of the dynamic process of adding one sample to the existing set each time, starting from a set consisting of the first sample. Therefore, totally 9 (97) generalized medians have effectively been computed for digits 1/2/3 (6). Taking the actual computation time for each update step of the whole

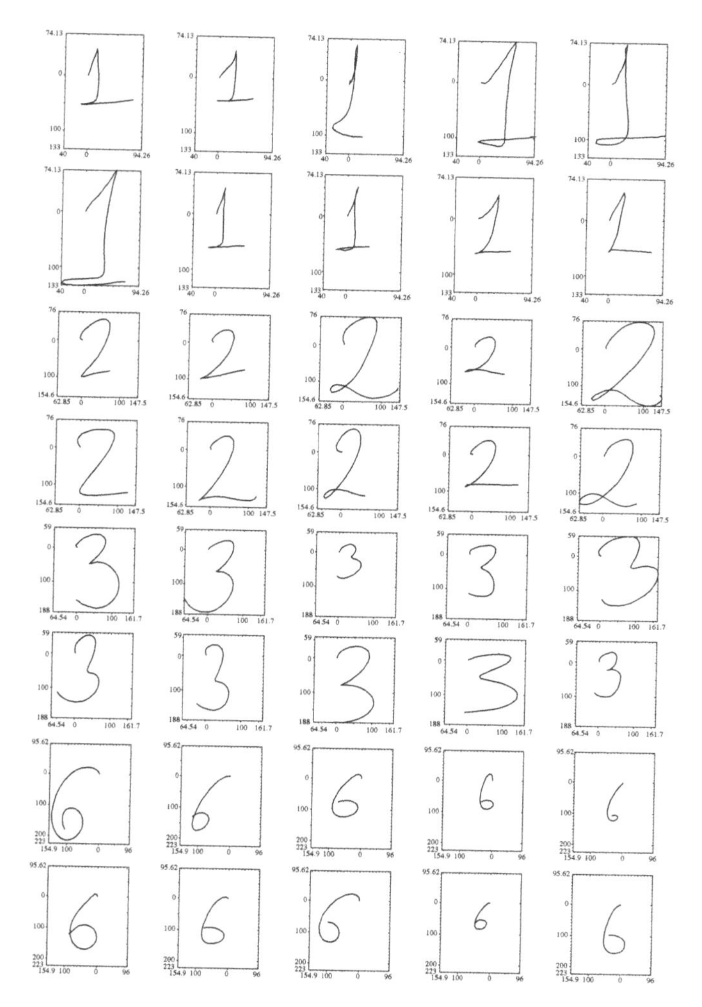

Fig. 2. Samples of digits 1, 2, 3 and 6, respectively.

dynamic process into account, the dynamic algorithm is attractive compared to static ones such as the greedy algorithm. If we use the static greedy algorithm to handle the same dynamic process, then the computation for digit 3, for instance, would require 46.10 seconds in contrast to 22.8 seconds of the dynamic approach. Considering the computational efficiency alone, set median is the best among the compared methods. But the generalized median string is a more precise abstraction of the given set of strings with a smaller consensus error.

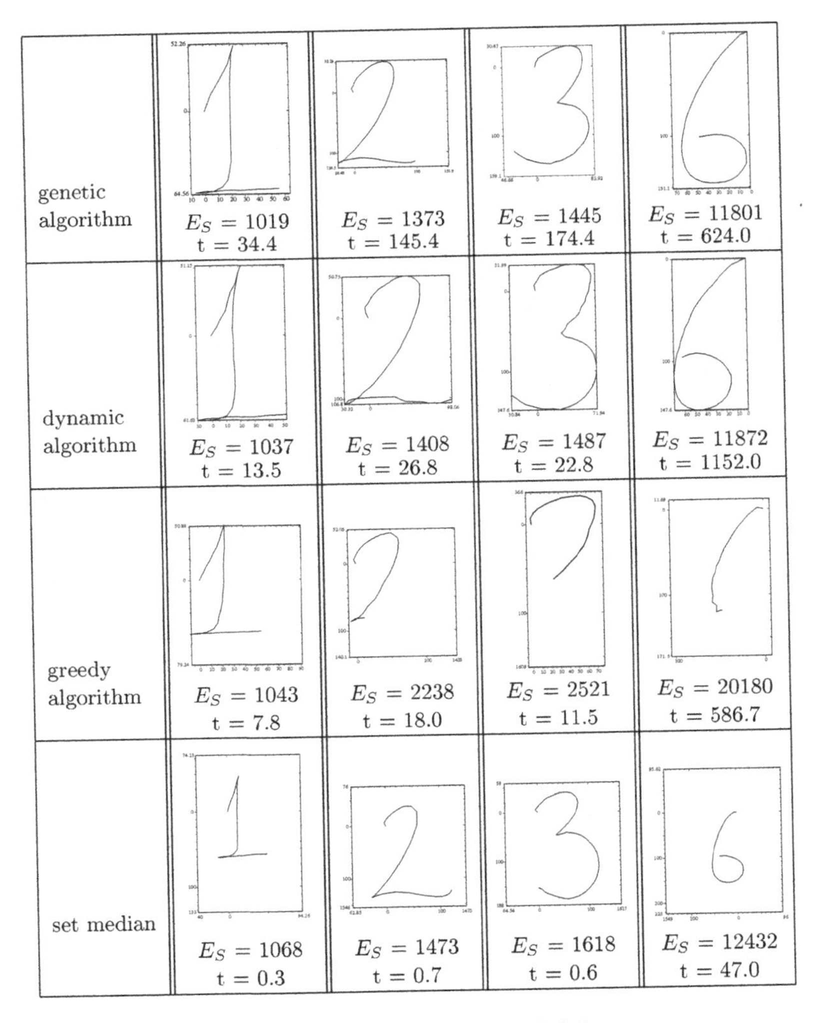


Fig. 3. Median computation of digits.

## 7. Discussions and Conclusion

In this paper we have reviewed the concept of median string. Particularly, we have briefly described several procedures for computing median strings. Experimental results were reported to demonstrate the median concept in dealing with a special case of time series, namely online handwritten digits, and to compare some of the discussed algorithms.

Table 1. Characteristic of median computation algorithms.

| Algorithm | String distance (original paper) | Extension to arbitrary distance | Handling weighted median | Handling center median |
|---|---|---|---|---|
| Exact algorithm and its variants [21,24,25] | Levenshtein | No | Yes | No |
| Greedy algorithms [4,22] | Levenshtein | Yes | Yes | Yes |
| Genetic algorithm [17] | Levenshtein | Yes | Yes | Yes |
| Dynamic algorithm [18] | Levenshtein | No | Yes | No |
| perturbation-based iterative refinement [20,26] | Arbitrary distance | N/A | Yes | Yes |

The majority of the algorithms described in this paper are based on the Levenshtein edit distance. The algorithms' applicability to an arbitrary string distance function is summarized in Table 1. Note that an extension to an arbitrary string distance function usually means a computational complexity different from that for the Levenshtein edit distance.

In the definition of median string, all the input strings have a uniform weight of one. If necessary, this basic definition can be easily extended to *weighted median string* to model the situation where each string has an individual importance, confidence, etc. Given the weights $w_q, q \in S$, the weighted generalized median string is simply

$$\bar{p} = \arg \min_{p \in U} \sum_{q=S} w_q \cdot d(p,q).$$

All the computational procedures discussed before can be modified to handle this extension in a straightforward manner.

The generalized median string represents one way of capturing the essential characteristics of a set of strings. There do exist other possibilities. One example is the so-called *center string* [15] defined by:

$$p^* = \arg \min_{p \in U} \max_{q \in S} d(p,q).$$

It is important to note that the same term is used in [13] to denote the set median string. Under the two conditions given in Section 3, it is proved in [15] that computing the center string is NP-hard. Another result is given in [7] where the NP-hardness of the center string problem is proved for the special case of a binary alphabet (i.e., $\Sigma = \{0,1\}$) and the Hamming string distance. The ability of the algorithms to compute the center string is summarized in Table 1.

Another issue of general interest is concerned with cyclic strings. Several methods have been proposed to efficiently compute the Levenshtein distance of cyclic strings [10,11,29]. It remains an open problem to determine medians of this kind of strings.


## Acknowledgments

The authors would like to thank K. Abegglen for her contribution to parts of the chapter.


## References


1. Apostolico and Galil, Z. (1997). (eds.), *Pattern Matching Algorithms*, Oxford University Press.
2. Bunke, H. (1992). Recent Advances in String Matching, in H. Bunke (ed.). *Advances in Structural and Syntactic Pattern Recognition*, World Scientific, pp. 3–21.
3. Bunke, H., Jiang, X., Abegglen, K., and Kandel, A. (2002). On the Weighted Mean of a Pair of Strings. *Pattern Analysis and Applications*, **5**(1), 23–30.
4. Casacuberta, F. and de Antoni, M.D. (1997). A Greedy Algorithm for Computing Approximate Median Strings. *Proc. of National Symposium on Pattern Recognition and Image Analysis*, pp. 193–198, Barcelona, Spain.
5. Crochemore, M. and Rytter, W. (1994). *Text Algorithms*, Oxford University Press.
6. Fagin, R. and Stockmeyer, L. (1998). Relaxing the Triangle Inequality in Pattern Matching. *Int. Journal on Computer Vision*, **28**(3), 219–231.
7. Frances, M. and Litman, A. (1997). On Covering Problems of Codes. *Theory of Computing Systems*, **30**(2), 113–119.
8. Fred, A.L.N. and Leitão, J.M.N. (1998). A Comparative Study of String Dissimilarity Measures in Structural Clustering. *Proc. of Int. Conf. on Document Analysis and Recognition*, pp. 385–394.
9. Gramkow, C. (2001). On Averaging Rotations. *Int. Journal on Computer Vision*, **42**(1/2), 7–16.
10. Gregor, J. and Thomason, M.G. (1993). Dynamic Programming Alignment of Sequences Representing Cyclic Patterns. *IEEE Trans. on Pattern Analysis and Machine Intelligence*, **15**(2), 129–135.
11. Gregor, J. and Thomason, M.G. (1996). Efficient Dynamic Programming Alignment of Cyclic Strings by Shift Elimination. *Pattern Recognition*, **29**(7), 1179–1185.
12. Guimond, A., Meunier, J., and Thirion, J.-P. (2000). Average Brain Models: A Convergence Study. *Computer Vision and Image Understanding*, **77**(2), 192–210.
13. Gusfield, D. (1997). *Algorithms on Strings, Trees, and Sequences: Computer Science and Computational Biology*, Cambridge University Press.

14. Guyon, Schomaker, L., Plamondon, R., Liberman, M. and Janet, S. (1994). UNIPEN Project on On-Line Exchange and Recognizer Benchmarks. *Proc. of 12th Int. Conf. on Pattern Recognition*, pp. 29–33.
15. de la Higuera, C. and Casacuberta, F. (2000). Topology of Strings: Median String is NP-Complete. *Theoretical Computer Science*, **230**(1/2), 39–48.
16. Jiang, X., Schiffmann, L., and Bunke, H. (2000). Computation of Median Shapes. *Proc. of 4th. Asian Conf. on Computer Vision*, pp. 300–305, Taipei.
17. Jiang, X., Abegglen, K., and Bunke, H. (2001). Genetic Computation of Generalized Median Strings. (Submitted for publication.)
18. Jiang, X., Abegglen, K., Bunke, H., and Csirik, J. (2002). Dynamic Computation of Generalized Median Strings, Pattern Analysis and Applications. (Accepted for publication.)
19. Juan and Vidal, E. (1998). Fast Median Search in Metric Spaces, in A. Amin and D. Dori (eds.). *Advances in Pattern Recognition*, pp. 905–912, Springer-Verlag.
20. Kohonen, T. (1985). Median Strings. *Pattern Recognition Letters*, **3**, 309–313.
21. Kruskal, J.B. (1983). An Overview of Sequence Comparison: Time Warps, String Edits, and Macromolecules. *SIAM Reviews*, **25**(2), 201–237.
22. Kruzslicz, F. (1999) Improved Greedy Algorithm for Computing Approximate Median Strings. *Acta Cybernetica*, **14**, 331–339.
23. Lewis, T., Owens, R., and Baddeley, A. (1999). Averaging Feature Maps. *Pattern Recognition*, **32**(9), 1615–1630.
24. Lopresti, D. and Zhou, J. (1997). Using Consensus Sequence Voting to Correct OCR Errors. *Computer Vision and Image Understanding*, **67**(1), 39–47.
25. Marti, V. and Bunke, H. (2001). Use of Positional Information in Sequence Alignment for Multiple Classifier Combination, in J. Kittler and F. Roli (eds.). *Multiple Classifier Combination*, pp. 388–398, Springer-Verlag.
26. Martinez-Hinarejos, C.D., Juan, A., and Casacuberta, F. (2000). Use of Median String for Classification. *Proc. of Int. Conf. on Pattern Recognition*, pp. 907–910.
27. Martinez-Hinarejos, C.D. Juan, A. and Casacuberta, F. (2001). Improving Classification Using Median String and NN Rules. *Proc. of National Symposium on Pattern Recognition and Image Analysis.*
28. Marzal, A. and Vidal, E. (1993). Computation of Normalized Edit Distance and Applications. *IEEE Trans. on PAMI*, **15**(9), 926–932.
29. Marzal, A. and Barrachina, S. (2000). Speeding up the Computation of the Edit Distance for Cyclic Strings. *Proc. of Int. Conf. on Pattern Recognition*, **2**, pp. 895–898.
30. Mico, L. and Oncina, J. (2001). An Approximate Median Search Algorithm in Non-Metric Spaces. *Pattern Recognition Letters*, **22**(10), 1145–1151.
31. O'Toole, A.J., Price, T., Vetter, T., Barlett, J.C., and Blanz, V. (1999). 3D Shape and 2D Surface Textures of Human Faces: The Role of "Averages" in Attractiveness and Age. *Image and Vision Computing*, **18**(1), 9–19.
32. Pennec, X. and Ayache, N. (1998). Uniform Distribution, Distance and Expectation Problems for Geometric Features Processing, *Journal of Mathematical Imaging and Vision*, **9**(1), 49–67.

33. Sanchez, G., Llados, J., and Tombre, K. (2002). A Mean String Algorithm to Compute the Average Among a Set of 2D Shapes. *Pattern Recognition Letters*, **23**(1–3), 203–213.
34. Sankoff, D. and Kruskal, J.B. (1983). (eds.). *Time Warps, String Edits, and Macromolecules: The Theory and Practice of String Comparison*, Addison-Wesley.
35. Sim, J.S. and Park, K. (2001). The Consensus String Problem for a Metric is NP-Complete. *Journal of Discrete Algorithms*, **2**(1).
36. Stephen, G.A. (1994). *String Searching Algorithms*, World Scientific.
37. Subramanyan, K. and Dean, D. (2000). A Procedure to Average 3D Anatomical Structures. *Medical Image Analysis*, **4**(4), 317–334.
38. Wagner, R.A. and Fischer, M.J. (1974). The String-to-String Correction Problem. *Journal of the ACM*, **21**(1), 168–173.
39. Wang, L. and Jiang, Y. (1994). On the Complexity of Multiple Sequence Alignment. *Journal of Computational Biology*, **1**(4), 337–348.